Praise

JONATHAN PINNOCK

LAST CHANCE
IN VEGAS

A MATHEMATICAL MYSTERY,
BOOK FIVE

This edition published in 2022 by Farrago,
an imprint of Duckworth Books Ltd
1 Golden Court, Richmond TW9 1EU, United Kingdom
www.farragobooks.com

Copyright © Jonathan Pinnock 2022

Print ISBN: 978-1-78842-429-5
Ebook ISBN: 978-1-78842-430-1

To Gail, as always

Author's Note

This is the fifth book in this sequence and a surprisingly large volume of adventurous water has flowed under the mathematical bridge since we first met Tom Winscombe. Being a reasonably conscientious sort of person, Tom does his best to paraphrase what has happened in the previous books where necessary for the benefit of new joiners. However, for readers who may require a little more in-depth background knowledge, I have put together an ever-expanding Wiki called Archiepyedia for this very purpose. You can find it on my website at www.jonathanpinnock. com/wiki_archiepye/. And don't worry, it has special buttons to protect you from spoilers!

Chapter 1

The long winter that followed my ignominious return from Belarus came and went, and in early March I was still living with my father in his static caravan just north of London. I would be wrong if I were to assert that this lifestyle was entirely without personal benefit – for one thing, I wasn't, for several very good reasons, paying any rent – but it would also be incorrect if I were to give the impression that I was one hundred per cent happy with the way things seemed to be turning out for me. The fact that I was pretty much entirely to blame for this state of affairs was no help either. I was, after all, the one who had accidentally incinerated the Vavasor papers, which my girlfriend Dorothy Chan had set her heart on acquiring. My *ex*-girlfriend Dorothy Chan, that is.

I missed her and I wondered if I would ever see her again.

It had been several weeks since I'd had any sort of conversation with anyone outside my immediate circle and the quality of communication within that circle was significantly limited, given that its members consisted of my father plus Wally the dog, μ the cat and Dolores and Steven the alpacas. Talking with my father could be a bit of a roller-coaster ride, but at least with the animals you knew where you stood most of the time.

'Morning,' I said to Dolores and Steven. There was a brief glimmer of recognition from each of them in turn before the realisation that I was carrying a couple of feed buckets kicked in and I was no longer the focus of their interest. 'Get your laughing gear round that,' I said as I put the feed down in front of them. Both animals plunged their heads in without acknowledging my presence any further and I retired to a safe distance. You don't want to disturb an alpaca when it's demolishing its breakfast.

I closed the gate to the field and headed back towards the caravan site. The daily routine with the alpacas was tiresome but at least it got me out into the fresh air at a reasonably early hour. I was going to miss them when they finally left us, although that all depended on Margot Evercreech, in whose care they had resided until I had accidentally stolen them. However, Margot was currently immersed, full-time, in sorting out her vast archive relating to the Vavasor twins, which had shifted state from being merely horrendously disorganised into completely and utterly chaotic following the previous year's visit to her cottage and subsequent ransacking by a deputation from the Fractal Monks.

Thinking about Margot Evercreech made me wonder what had happened to Benjamin Unsworth, who I'd last heard of impersonating the chaotician Dr Rory Milford in order to infiltrate the Belarusian Petrov mafia organisation on behalf of Helen Matheson, the ex-spy. This was after my own, failed, attempt to impersonate Dr Milford, but before the unexpected arrival of the real one. I hoped that Benjamin had managed to get out of Minsk alive. He was clueless but he meant well and I would have hated for anything to have happened to him, not least because if I hadn't cocked up my own mission by getting kidnapped at the airport by the rival Gretzky gang, he wouldn't have even been sent there. I'd tried calling him a few times, but his mobile was never on. I hoped this just meant that he'd abandoned it to avoid being tracked.

Thinking about Benjamin Unsworth and everything that had happened in Belarus did at least have the effect of putting things into perspective. Since moving in with my father, I hadn't had to throw myself out of the top of a burning building once, for example. So that was all good. But I couldn't carry on this way forever. At some point I had to reconnect with the rest of the world. I was still young. There were still things I had to do.

As I approached my father's pitch, my phone went off. The number looked vaguely familiar, although I hadn't reconstructed my address book since losing my last phone somewhere in the middle of a highway in Belarus.

'Hello?' I said.

'Hi, fuckwit,' came a familiar female voice.

'Ali?'

'Yeah, the same. Listen, Winscombe, you owe me one.' Yes, it was definitely Ali. And, to be fair, she was right. Last time I'd spoken to her, she'd helped me find the parachute that enabled me to escape from the top floor of that burning building in Minsk. However, owing to my subsequent estrangement from Dorothy, Ali's business partner, I hadn't really had the opportunity to properly thank her for what she'd done.

'Ah,' I said. 'Yes, about that. I really meant to—'

'Oh, shut the fuck up and listen, twatface.'

It was good to be talking to someone other than the animals and my father for a change. Ali wouldn't have been my first choice, mind, but I wasn't in any position to be choosy.

'OK, I'm listening.'

'Good. Right. Well.' This was unusual. Ali almost seemed nervous. I'd never known her like this before.

'Go on,' I said.

'Yeah well, this is kind of awkward.'

I suddenly thought of something. 'Is it about Dorothy?'

Ali seemed genuinely surprised. 'Dot? Fuck, no.' She paused. 'Look, the thing is, Patty and I are going to have a baby.'

'What?' Given that the reason why μ had initially moved into the care of my father was that Dorothy and I hadn't trusted Ali to look after a small animal, the thought of her being responsible for a small human being was challenging, to say the very least.

'We're going to have a fucking baby, Winscombe.'

'Right,' I said. 'Well, congratulations, I guess.' Thinking about it, Ali's partner Patrice was more than capable for compensating for whatever Ali lacked in the caring stakes, so perhaps I had jumped the gun a little. 'So when's it due?' I said, keen to show interest.

'Well, a little over nine months.'

'Oh,' I said. 'Sorry?'

'Oh come on, Winscombe. Work it out. Fuck's sake.'

I thought about it for a moment, but it still didn't make sense.

'Christ almighty, Winscombe,' said Ali. 'Did your poor benighted mother not have that little chat with you?' There was a deep sigh from her. 'Look, the thing is, we haven't actually conceived yet. We're, um, short of one important component. And I would like to say, at this point, that this was very much not my idea at all, in fact I was strongly against it and, if it hadn't been for the fact that I couldn't come up with any alternative suggestions I wouldn't be asking at all but Patty has somehow got it into her head that, despite being an utter fuckwit, you're some kind of survivor and *that's* what matters to her most importantly – and it's not as if you'll have any ongoing involvement so it's not tying us to any sort of long-term commitment and so that's it.'

There was a long silence.

'Are you asking me to be the father?' I said eventually.

'Basically, yes. Fuck knows why, but here we are.' I could hear her briefly putting her hand over the microphone at her end and shouting 'Yes, I'm talking to the twat now. Stop hassling me!'

'Gosh,' I said. I genuinely didn't know what to say. 'So, which one of you is going to be—'

This was apparently the funniest joke she'd heard all year. 'Fucking hell, Winscombe. Do you seriously see me doing the job?'

She was right. It was, quite literally, inconceivable.

'No,' she continued. 'I may have all the appropriate organs somewhere inside me but I have no intention of making the slightest use of them. The idea of having some kind of weird alien life form growing in there for the best part of a year makes me want to fucking vomit.' She paused. 'No, Patty's doing the honours and – before you get any ideas – she's going to take it bottled, not draught.'

'The thought never crossed my mind,' I said.

'Yeah, right. I've met your type before.'

'What type's that?'

'Men.'

'Hey, not all... yeah, OK. OK.' I wasn't going to go there. 'I feel honoured,' I said instead.

'Well, don't. Just come round tomorrow evening and Patty'll go through the logistics. I take it you're not busy?'

'I can probably fit you in. Look, have you heard anything from—'

'See you tomorrow, then,' she said. 'Jesus,' she added, 'I can't believe I've had to do this.' Then she hung up.

I stood there for a good twenty seconds staring at my phone, wondering how much of the preceding couple of minutes was a figment of my imagination. Then I shook my head and carried on walking.

My father was sitting in his pyjamas on the steps of his static caravan, smoking a roll-up cigarette. He held up a hand in greeting as I approached.

'Morning, son!' he managed to gasp out before he was racked by a fit of coughing that would have despatched many a lesser man to his grave.

'Morning, Dad,' I said. 'You really should try to cut down.'

'Only thing that gets me going in the morning, this,' he said, shaking his head and taking a deep drag. 'If I didn't smoke, son, I'd be dead.'

I had long since abandoned any attempt at questioning my father's idiosyncratic approach to self-medication, so I didn't pursue the matter further. He had always made poor lifestyle choices and there was no way he was going to change now. Besides, he used to say, why was I so worried about him, given what was due to me when he eventually did pop his clogs? After all, one day all this would be mine.

I stepped past him and walked into my inheritance. Wally the dog bounded over and gave my crotch a welcoming slobber. Satisfied, he let off a silent fart before turning round and ambling back to his moth-eaten bed in the opposite corner. Meanwhile μ had emerged and was giving me her usual disappointed look that left me in no doubt that she had still not forgiven me for bringing her to this godforsaken place and probably never would, as long as she lived.

I got a bowl out of the cupboard, filled it up with cornflakes and went to the fridge for milk. I gave the bottle a sniff and quickly decided that I would be eating my breakfast cereal dry today. Fortunately, the packet had been left open, so they were a little on the soft side already.

'Milk's off,' I said as my father came back in.

'And who's turn was it this week?' he said, taking a seat opposite me.

'Yours, Dad.'

'Pfft.' The lack of milk wasn't much of a problem to him as he didn't eat breakfast anyway most days. Besides, I very much doubted that he had much left in the way of functioning sensory organs so, even if he did, he probably wouldn't have noticed anything wrong with the milk. He sat back and folded his arms, deep in thought. Then he scratched his head and pointed at me in a vaguely accusatory manner. 'Your problem,' he said, 'is that you don't have any purpose these days.'

There were a number of ways in which I could have countered this. I could have pointed out that, as far as I was aware, there had never been anything remotely close to such a thing in his own life, so why should he expect it in his offspring's. But I wasn't in the mood for that kind of argument this early in the morning. Alternatively, I could have mentioned my recent phone call with Ali, but that would have led to all manner of complex explanations which, again, I didn't feel up to yet. Instead I just shrugged and let him have his parental teaching moment.

'Y'see? You know I'm right, son,' he said. 'Maybe it's about time you looked for a job again. I can't keep on supporting you at my age.'

'Dad, you're not supporting me,' I began, but then realised very quickly that any detailed explanation of my own financial arrangements might also get quite complicated very quickly – given that I was still living off the proceeds of hot laundered money liberated by Dorothy and myself from the Autonomous Bailiwick of Channellia cryptocurrency rig the previous summer, just before it collapsed into the sea off Burnham in Somerset, topped up with the remains of a wad of Belarusian roubles given to me by Helen Matheson for my mission in Minsk.

The annoying thing was that the old man was right, even if for completely misguided reasons. I needed to sort my life out. I needed to find a purpose. For a brief time, when I had been with Dorothy, everything had seemed to make some kind of sense. But now I was cast adrift on a sea of confusion with neither compass nor effective means of propulsion.

There was only one way I was ever going to get Dorothy back, and that was never going to happen. There had only ever been one copy of the Vavasor papers and I had been the one who had turned them to a smouldering heap of ash. There was absolutely no way that I would ever be able to bring those papers back into existence and that was the only thing that ever mattered to her. Well, that and the fact that she was also convinced that I'd been carrying on with Helen Matheson, although I'm sure I could have sorted that out, given enough time. But time was one thing that Dorothy was unlikely to be giving me ever again.

'Maybe you're right,' I said. 'Maybe I do need to find some direction in life.'

My father nodded sagely and tapped his head. 'Still some old wisdom locked away up here, son,' he said.

'Yeah well, let's not get carried away.'

'How are the llamas this morning, anyway?' he said.

'Dad, they're alpacas.'

'Same difference.'

'I'm sure they wouldn't see it that way,' I said. 'Anyway, they're fine. Bright and perky as ever.'

'Good,' he said. 'Now. Need to make some calls and, um, check my portfolio. My phone's out of battery, so can I borrow yours?'

This happened quite frequently, usually coinciding with my father running off the end of his pay-as-you-go data allowance.

'Only if you don't download any more dodgy content onto it,' I said.

'I don't know what you mean.'

'Last time you borrowed it, it came back with half a dozen cryptocurrency trading apps. We had a conversation about this, didn't we? No more crypto, Dad.'

'But we did all right out of those – what do you call them? – Tulpencoins, didn't we?' he said. Tulpencoins were the cryptocurrency that the Channellia bunch had been promoting as part of a complex scam. Most of the people involved with that, along with the servers that controlled it, had ended up at the bottom of the Bristol Channel, although the account I'd given my father of what had happened on board Channellia had been heavily filleted in order to avoid triggering his impressively welcoming attitude towards mad conspiracy theories.

'The only reason you made anything at all out of that scam was the very generous compensation scheme,' I said. There was, in fact, no compensation scheme other than the one that Dorothy and I had put together ourselves purely for him. This was the cover story for slipping him some of the laundered cash in return for a cast iron promise not to get involved in anything like this ever again.

'But One-Eyed Kev—'

'As far as I know, your neighbour, One-Eyed Kev, is not regulated by the Financial Conduct Authority, so he is in no position to be offering you advice. Besides, he's an idiot.'

'All right then,' said my father with a sigh. 'No more crypto.'

'Good,' I said, tossing him the phone. 'And don't get it covered in mud again either.'

As he wandered off, I looked at Wally and I swear he gave me an eyeroll. Then again, he'd lived with my father longer than I had, so he really ought to have been inured to this kind of thing. Would I be any better as a parent, though? I'd never had much cause to consider this before, but now that I'd been put in the frame for the role – albeit in the most minimal way possible – the idea suddenly became a lot more concrete.

What did being a parent actually involve? There were a thousand different aspects to it, but maybe it all boiled down to doing whatever you could to ensure that your children grew up feeling secure and happy. Which is probably why my confidence in my potential abilities in this particular field of human endeavour took something of a knock the very next morning, when I went to feed the alpacas and found that they were no longer there.

Chapter 2

I put down the buckets of feed and ran to the gate, which was swinging wildly on its hinges. Had some drunken idiot come along in the night and opened it for a lark? Had I forgotten to close it properly the previous evening when I'd come out to check up on them? I ran through it in my mind but I was convinced that I'd double and triple checked like I usually did before leaving them.

What was I going to say to Margot Evercreech? 'You know those alpacas I borrowed from you? Well, looks like you might have to wait even longer until you get them back. No, they're fine. Really, they're just fine. Wherever they are, I'm sure they're fine. Sorry? Can't hear you, you're breaking up.'

Oh god, what was I going to do?

I took a deep breath. If they had escaped, they surely couldn't have gone far. The first thing to do was put up a notice on every lamp post in the vicinity: 'MISSING! Dolores and Steven. Alpacas. One black, one white. Fluffy bodies and fluffier brains. Tend to act a bit stoned.' Then again, the more I thought about this, the more pointless that seemed. I wasn't actually sure how fast they could go, and I suspected they could really motor if they put their minds to it. Moreover, for all I knew, they'd been out all night – they could be halfway to the M25 by now.

It wasn't as if they were some dozy cat that had accidentally got locked in a neighbour's shed.

I took another deep breath. Come on, I'd got out of worse scrapes than this. All I needed to do was find another couple of alpacas that looked roughly like Dolores and Steven, and substitute them. But the problem with that was that I hadn't actually taken many pictures of them and there was every chance that Margot would spot a missing dappled mark on Steven's left buttock or the fact that Dolores's neck was a couple of centimetres too long.

No, I had to find them.

I took a third deep breath.

Maybe there were security cameras somewhere along the route here that I could get the police to look at, to see if there were any sightings of rogue alpacas. But I really didn't want to get the police involved in this, owing to the fact that my acquisition of the two animals hadn't been entirely legal – even if I could perhaps argue that I had been forced to take them with me while escaping from being kidnapped by their owner. But anything that involved an explanation of more than half a dozen sentences ran the risk of making me look a madman.

There had to be clues.

Looking at the ground, I noticed a pair of tyre tracks for the first time, leading from the gate into the field and back out again. So there it was. There was no point in printing any flyers for lamp posts. Dolores and Steven had been stolen.

What kind of maniac went around stealing alpacas, though? Was camelid-rustling a thing in the home counties? Did someone sidle up to you in the pub and say, 'Slip us a monkey and I can do you a nice pair of hot llamas, no questions asked'? The very idea was offensive, although it was also entirely possible that both Dolores and Steven were well on their way to joining the food chain by now.

I stood in the field for several more minutes, willing them to rematerialise from whatever alternative reality they had temporarily wandered into, but it was no good. However hard I tried, the field remained empty apart from my own increasingly desperate self. I gave up and walked back to the gate, slamming it closed for no good reason before stomping back to the caravan.

'The alpacas are gone,' I said to my father.

'What?' he said, looking up from the table where he was carefully constructing his first roll-up of the day.

'The alpacas. Dolores and Steven. Gone.'

'What do you mean, gone?'

'Someone's kidnapped them, Dad.'

'Why?'

'I've no idea, Dad. Have *you* any idea why? Because I'd love to know.'

'Hey, calm down, son. They've probably just nipped out.'

'Nipped out? Like, they found they were out of Weetabix or something and popped round to the corner shop?'

'No, not like that—'

'Because if that's the case, why did *both* of them go, eh? Eh? Or maybe Dolores said to Steven, "You know what, I fancy stretching my legs, so I might as well come along too"? Is that it? Because I don't think that's how alpacas work, Dad.'

'Now don't you get all umpty, son.'

'I'll get as umpty as I like. I've just lost two alpacas that I was supposed to be looking after and you don't seem to be remotely interested.'

'They'll come back, son. Trust me. I know these things.'

'You know sod all about anything, Dad.'

My father didn't respond to this but picked up his manky roll-up and embarked upon an increasingly irritating sequence of failed attempts to get his lighter to ignite.

'Here,' I said, grabbing a box of matches from the shelf next to me. I struck one and held it up to him. He allowed me to light the end of his cigarette, then leaned back and drew deeply on it. Then he exhaled and stared up at the ceiling.

'Son,' he said. 'Don't take this the wrong way, but I think that, of late, you have perhaps made too much of an emotional investment in those llama things.'

'They're alpacas, Dad.'

'Whatever.'

My father was, generally speaking, fond of animals but he had never really taken to Dolores and Steven. Until I'd begun my temporary sojourn with him, he'd left the feeding and general maintenance of them to his neighbour, 'Mad Dog' McFish.

'Hold on,' I said. 'Are you basically saying that we shouldn't care what's happened to them?'

He shrugged. 'Well they weren't going to be here forever, son. You were always going to have to say goodbye at some point.'

'I would have liked to have handed them over properly, though. It would have been nice to feel that I'd done my bit in looking after them. Just for once in my life, I'd like to get something right in a nice, uncomplicated way.'

'If I had a pound for—' began my father.

'No,' I said, thumping the table. 'No folksy wisdom today, Dad. That's it, I've decided. I'm going to get those alpacas back.'

Stirring music crescendoed towards an epic orchestral climax in my head as I stormed out, brought to an abrupt *cesura* as μ materialised from nowhere at the top of the steps leading out of the caravan, causing me to take an elaborate experiment in evasive action that eventually resulted in me finding myself face down on the ground at the bottom. After a moment or two, I began to feel gentle drops of rain on the back of my neck.

'Bollocks,' I said. I turned to look back at μ, who was staring down at me as if deciding whether or not to give me an extra half point for attempting a high-tariff element in a gymnastics routine. By the time I got up and brushed myself down, she had got bored and had gone back into the caravan out of the rain, which was coming down quite heavily now.

I trudged back to the alpacas' field to see if I'd missed any clues. Perhaps whoever had taken Dolores and Steven had dropped something from their pocket? Thinking about it more logically now, I realised that the alpacanappers – depending on how many there were – would have had to get out of their vehicle at least once and maybe twice if there were just one: to open the gate and to herd the animals into their van. I squatted down on my haunches at the entrance to the field, but there was nothing there at all.

However, when I walked to the end of the tyre tracks in the middle of the field, I spotted something that I'd previously missed: something yellow, floating in a puddle that was forming a few feet away as a result of the rain. I bent down and picked it up, shaking the drips off. It was a rubber duck and attached to it was a key.

'Hello?' I said, bending down to speak into the entry phone at the bottom of Patrice's block of flats near the Willesden High Road. 'It's me, Tom.' I almost added 'The Daddy,' but realised just in time that it might have sounded a bit crass. Besides, I wasn't feeling particularly pleased with myself, having spent most of the day fretting about what might have become of Dolores and Steven, so it probably wouldn't have come out right. The lock buzzed and I pushed the door open. I went up the stairs to find Patrice waiting for me with her arms wide open and a big beaming smile on her face.

'Thomas!' she said, drawing me close for a hug. 'So good to see you again!'

'It's lovely to see you too,' I said. 'Been a while.' I was feeling better already.

I followed her into the sitting room. She'd had the floors sanded since the last time I'd been there and walls on two sides were now completely lined with shelves, all neatly filled with arrays of books that gave off a strong whiff of academia. Patrice was a professor of mathematical biology and she wore her learning lightly, but occasionally you got a blast of it full in the face.

There wasn't much evidence that Ali was also living here, apart from her presence in the room. She was lounging in one of the armchairs and staring up at the ceiling. Instead of her usual Doc Martens, she was wearing fluffy slippers with bunny heads on the toes and, not for the first time, I marvelled at the concessions she was prepared to make for Patrice. I don't think I'd ever encountered anything like this before when it came to Ali and it was truly a wonder to behold.

I sat down on the end of the sofa nearest her. It was a sensible, rational sofa, offering just the right balance between comfort and support and I felt instantly more at home than I'd felt in the entire time I'd spent at my father's. Patrice took the other end.

'Hi, Ali,' I said. She didn't reply.

'Alison,' said Patrice, in the kind of tone that was usually reserved for a nine-year-old child who had just been caught supergluing their classmate's ponytail to the back of their chair.

'Oh, hi there, Donor Boy,' said Ali, her voice a supersaturated solution of finest quality sarcasm.

'Alison, would you like to offer our guest a drink?'

'A drink?' said Ali, evidently flabbergasted that she should be required to be in any way hospitable to this intruder.

'Yes, Alison. A drink.'

'Jesus. OK, Winscombe,' said Ali, turning to me. 'What do you want?'

'Um… beer?' I said.

'Beer.'

'From the fridge,' said Patrice.

'Jesus.' Ali got up and stomped off towards the kitchen.

'She doesn't mean it really,' said Patrice.

I rather felt that she did, but I didn't say as much.

'Thank you so much for coming over,' she continued. 'I think Alison has explained what we're looking for?'

'Basically, yes. I'm honoured. I think.'

'Well, cometh the hour, cometh the man, Thomas. The thing is, Alison and I can probably do enough between us by means of either nature or nurture to bring up an intelligent, inquisitive child, but we – or at least *I* at the moment, but I think Alison will come around eventually – feel that you have a different quality to add to the mix.'

This was certainly true, although I still wasn't sure that whatever I did have was going to be of much use to them.

'You see, the thing is, Thomas, you're a survivor. Somehow, despite whatever crazy, scary, unpleasant stuff gets thrown at you, you seem to come out of it smelling of roses. Even Dorothy—' She checked herself, as if she'd said more than she'd intended to '—well, anyway, that's what we – I – think and that's why we've asked you to come over to talk about it.'

Ali appeared again, holding two beer bottles. She thrust one in my direction, took the other one and went back to her armchair, draping herself inelegantly back over it.

'Cheers,' I said, taking a swig. It was warm.

'I'm not sure you should even be drinking that, pal,' said Ali. 'Alcohol isn't good for your product.' I suddenly felt guilty. Perhaps I wasn't taking this as seriously as I should.

'Well, it's not as if Thomas is going to be donating anything quite yet,' said Patrice, smiling.

'Well, that's a relief,' I said. 'And my responsibilities will end… when?'

'As soon as you've jizzed into the fucking test tube,' said Ali.

'Right,' I said. 'Right.'

'Just don't expect me to shake your hand afterwards, that's all.'

'So, Thomas, are you happy to do this?' said Patrice, flashing a warning glance in Ali's direction.

'I guess so.'

'The people at the clinic will need to check that you haven't got any inheritable conditions,' said Patrice.

'Yeah, like rampant stupidity,' said Ali.

'Alison, please,' said Patrice. 'I know you're finding this difficult but try not to insult Thomas when he's trying to help us.'

'It's OK,' I said. 'I'm used to it. So is there some kind of contract I have to sign?'

'Ah, yes,' said Patrice. 'I suppose we need to do this properly.' She got up and left the room. She returned with a sheaf of paper stapled together, which she handed over to me. I scanned it to check that I wasn't signing over anything important and then added my name at the bottom and passed it back to her.

'Wonderful!' she said. 'I'll let them know at the clinic and they can arrange for you to come in and do your thing. Alison?' she added, turning to her partner. 'We're going to have a baby!'

'Fuckin' A,' said Ali. 'Just as long as the little bastard doesn't inherit his stupid dumb face, I'll be happy.'

'Cheers,' I said, taking a swig from my beer. There was a companionable silence for a minute or so.

'Well then,' said Patrice. 'What are you up to these days, Thomas?'

That was a very good question and one that I struggled for a moment to answer. 'I've been... taking it easy for a bit,' I said. 'You know, looking after my dad and the alpacas. That sort of thing.'

'Right,' said Patrice.

'Actually, speaking of the alpacas,' I said, 'there was something I wanted your advice on. You see, the thing is, they've been stolen.'

'Oh no,' said Patrice.

'Well, that was fucking careless,' said Ali.

'But they left this behind,' I said, extracting the key ring from my pocket.

'Also fucking careless,' said Ali, taking it from me and studying it. 'So I suppose you want me to track down where this came from?'

'Um, yes,' I said. 'Basically that. I remember the way you found Channellia from my old disc drive.'

'Yeah, well,' said Ali. 'That was a piece of live computer equipment. This is an inert piece of yellow plastic.'

'But you're good at that kind of thing.'

'Yeah, that's true. I am pretty fucking awesome.' She swung her legs off the armchair and slouched over to a desk in the corner, where she sat down, opened up a laptop and started tapping away at the keyboard.

Patrice got up. 'I think I might as well go and get the food ready. You are staying, Thomas?'

'Oh yes, please,' I said. It had been a while since I'd tasted Patrice's cooking and I had no intention of going back home without taking the opportunity to taste it again. Patrice left the room and for five or ten minutes there was just the sound of Ali's fingers on the keys of her computer.

'Getting anywhere?' I said eventually.

'Low-level interrupt received,' said Ali without looking up. 'CPU fully utilised, can't handle. One... two... three... oh, TIMEOUT. Bad luck. The interrupt has timed out and is being ignored.'

I took this to mean that I was to leave her alone. The thing about Ali was that once you gave her a challenge like this, she tended to forget that the challenge had been presented to her by someone she regarded as being on a par with one of those parasitic worms whose lifestyle revolves around invading a snail's eyestalks in order to get eaten by a passing bird. For the duration of the task, everything else became completely secondary. I decided that this was probably a good time to make myself scarce.

I wandered into the kitchen and was immediately beset by the heady scent of Bajan cuisine. Patrice was chopping up vegetables. There was a tank on the work surface next to her, from which a strange little creature, with what looked like the widest beaming smile and the tiniest eyes I'd ever seen, was peering out at me. It seemed to be the result of a deeply unethical genetic experiment at creating some kind of hybrid fish lizard.

'That's Axel,' said Patrice. 'Axel the axolotl.'

'Cool,' I said. 'Is he new?' Patrice had a large collection of slithery things and I wondered briefly how the planned addition to their family was going to cope with all this. Perfectly well, probably.

'Yes. Always wanted one of them. I think Alison is getting to like him too.'

I raised an eyebrow at this. Patrice just smiled. I had a feeling that she rather enjoyed pushing at the edges of her relationship with Ali just to see how far she could go, and I rather admired her for that. It was probably quite good for Ali, too.

'Anything I can do?' I said.

'Well, you can give that sauce a stir,' said Patrice. 'And then there's always the washing up.'

I agitated the sauce a little and then went over to the sink.

'So, how's Alison getting on?' she said.

'She wouldn't say.'

'That figures. She tends not to come up with interim reports.'

'That's certainly true.' I continued with the washing up for a minute or two. 'Patrice?' I said. 'Have you heard anything of—'

'Dorothy? I was wondering when you were going to ask.'

'I wasn't sure if I was allowed to ask. Ali seemed a bit cagey when I spoke to her on the phone before coming here.'

'Well, it's a difficult situation.'

'Yeah, well, I'm sorry if I—'

'Oh no, it's nothing to do with you. It's Third Uncle.'

'It's who?'

'You know, Third—'

She broke off as Ali appeared in the doorway.

'Right,' she said, folding her arms and leaning against the wall. In between the thumb and index finger of her right hand, she held the rubber duck. 'Here we go, kids.'

Chapter 3

'This key fob,' said Ali, 'was almost certainly manufactured by the, no doubt once very excellent but now pretty much defunct, Luxxy Duxxy corporation. And, yes, I'm sure you will be ecstatic to find out that the Luxxy Duxxy corporation was a toy manufacturer who used to produce nothing but novelty rubber ducks. You know the type of thing. Celebrities' heads on them? That one from *Bake Off*. That footballer with the mad hair. Josef Fucking Stalin, I dunno. Anyway, looks like they went under last year. Got sued into oblivion by some twat of a singer who'd gone and copyrighted the design of his moustache. The fine artefact that I'm holding in my hand here never actually went on sale.'

'So how come you know it was made by them?'

Ali's eyes lit up. 'Aha!' she said. 'Found an image of it in a PDF of their final company report. Looks like they were planning to diversify before they went under. I'm guessing this is a prototype that got left behind when they turned the lights off and went home.'

'Any idea where they're based?'

'According to their last set of accounts, they've got properties in Cumbernauld, Penzance and Basingstoke. Now the one in Basingstoke is interesting because it also turns up in a list of properties currently rented by Merritt Foods.'

'That sounds ominous,' I said. 'Who are they?'

'US outfit,' said Ali. 'Animal feeds. Headquarters in Little Rock, Arkansas. In the last few years, pivoting towards real estate and gambling with interests in Las Vegas and Atlantic City. Owned by one Robert J Merritt III. Billionaire and a massive investor in a number of our favourite right-wing causes, including – ta da! – the very wonderful Institute for Progress and Development.'

'Oh god,' said Patrice. 'Them again.'

'You're not their biggest fan, are you?' I said. I wasn't that much of a fan either. The IPD was one of those opaquely funded right-wing think tanks that was somehow involved in a significant number of fundamentally bad things, from the Tulpencoin cryptocurrency scam to the Belarusian financial market manipulation scheme.

'Well, as a lesbian academic of colour,' said Patrice, 'I'm pretty close to the heart of their Venn diagram of hate. The only way I could improve on that is if I turned out to be transgender. Sometimes I'm almost tempted, just to wind them up.' For a moment, the main sound in the kitchen was her knife attacking a cabbage with renewed vigour.

'But what would this Merritt guy want with a couple of random alpacas?' I said.

'Fuck knows,' said Ali. 'That's where I've drawn a blank. Can't imagine they'd be worth stealing to make pig nuts out of.'

I thought about this for a moment.

'I'm going to go there,' I said.

'Thomas,' said Patrice. 'You need to keep yourself safe. For us. If the Institute for Progress and Development are involved, it could be really dangerous.'

'I'll be fine,' I said. 'What's the worst that can happen?'

'You could get yourself killed, that's what could happen.'

'Oh, let him,' said Ali. 'If he wants to get murdered, that's his problem.'

'What about – you know – our family?' said Patrice.

'Ach, we'll find someone else. Twattish men are ten a penny.'

'Cheers,' I said.

'You're welcome,' said Ali. 'Here,' she said, handing me an old envelope that she'd scribbled something on. 'Luxxy Duxxy, Basingstoke. I've even given you the postcode.'

I turned towards the tank where Axel was watching the three of us with interest. 'What do you reckon?' I said. The little amphibian peered at me closely and then appeared to blink. I turned back to Patrice. 'Axel says go for it,' I said.

'Promise me you'll be careful, Thomas.'

'Of course. And you said it yourself, I'm a survivor.' I almost believed her too.

Ali snorted with derision. 'Survivor, my arse,' she said. 'The only reason you're still alive is because of all the other folk around you – mostly women – who've been waiting to catch you when you fell.'

This seemed unfair, although I had to admit there was an element of truth to it. 'That reminds me,' I said. 'What were you saying about Dorothy, Patrice?'

Patrice looked at Ali. Ali looked back at Patrice, then at me. 'You really want to know?' she said.

'I think so.'

'You think so.' Ali took a deep breath and grimaced. Then she sighed and sat down at the kitchen table, putting her head in her hands. She shook her head and looked up at Patrice. Patrice gave the slightest of nods. 'OK,' she said, 'here we go.'

'Look, if it's going to be difficult—' I began, but she ignored me.

'Once upon a time there were two programmers. One of them was, let's face it, probably one of the brightest and best on the fucking planet. That's me, by the way, but you probably knew that already. The other one was pretty hot too but she also had the commercial nous. Yeah, Dot had a great eye for what would sell. They made a great team and they built some great games. They fought off the trolls, kept the venture capitalist vultures at bay and were beginning to make something big and beautiful.

'Then some twat walks in and says to Dot, "Hey, you know that mystery about those fucking stupid Vavasor twins you used to be obsessed about? Guess what. I've just found their missing papers."'

'That's not quite how it happened,' I said.

'Might as fucking well be, pal,' said Ali. 'Anyway, madam drops everything to go and chase shadows, leaving me holding the baby – as it were,' she added, looking at Patrice. 'And from then on, nothing is ever quite the same.'

'But the Vavasor thing is over,' I said. 'We solved the mystery of how they died and… well, I burnt the papers.'

'Is it over, though? Trouble is, once you've chased one lot of shadows you get a taste for it and suddenly shadows are the most important thing in your life. Sitting at a desk, building the world's greatest new software empire is somehow not quite as exciting.'

'But where is she now?' I said.

'Macau,' said Ali.

'What?'

'Yeah, Macau. Remember what I told you about Third Uncle? Used to run the second-biggest casino out there? Well, she gets a call from him one day at work. Some kind of trouble going on, and instead of saying what any normal person would to a long-lost relative phoning up out of the blue from the other side of the world – which is basically "Tough titties, pal, sort your own shit out, I'm busy," – what does Dot say? "Ooh, that sounds exciting, I'll be there as soon as I can get a fucking flight."'

'Ah,' I said. 'But that's nothing to do with me, surely?'

'Wouldn't have happened in the pre-Winscombe era, pal.'

'But when was this? Have you heard from her since?'

'Month or so ago. And no, I haven't bothered calling her. I can run the fucking show without Dot. She'll crawl back one day and if the whole thing's gone down the tubes in the meantime, she's only got herself to blame.'

Her story apparently over, Ali leant back in her chair and stretched. 'Right, is food ready?' she said. 'I'm fucking starving, man.'

The following evening I found myself on the way to the Millennium Business Park on the outskirts of Basingstoke. The taxi driver from the station, instead of finding it a little odd that I was coming out here at past six o'clock on a cold early spring evening, simply said, 'I expect you're with the bunch at the old Luxxy Duxxy place, then?' so I knew I was on the right lines.

'So what do you lot get up to in there, mate?' he said as we approached the industrial estate.

'Well,' I said, 'I could tell you, but I'd have to kill you afterwards.' I followed this up with some performative nose-tapping, just to hammer the point in.

'Ha ha, I bet you would, mate.'

'Honestly, you wouldn't believe what goes on,' I said. There was no harm in getting a bit of local gossip going.

'I can imagine, mate. Anyway, here we are.'

I paid the fare and got out. The taxi disappeared into the distance. The Luxxy Duxxy unit was surrounded by high metal fencing with razor wire at the top. In the fading twilight I couldn't see any security cameras but I assumed there must be some hidden around the outside of the building, so I wrapped my scarf around my face just in case. The entrance gate was secured in place by a chain looped through a padlock that looked as if it would break my foot if it happened to drop on me. I tried my key in it, but it didn't fit. I gave it a good tug anyway, but it showed no interest whatsoever in opening. A sign attached to the gate announced that it was under 24-hour surveillance by Guardians of Basingstoke Ltd with a phone number to call if I saw anything suspicious. I made a note of it, just in case.

Well, this didn't look terribly promising. Peering through the fence, I could see that the building was a largely windowless concrete construction. A faded painted sign on the side of it declared it to be the home of Luxxy Duxxy, and there was a vast picture of a rubber duck with the head of David Beckham, for no good reason, attached to it. From where I was standing, I could only see one door, which appeared to be the fire exit. I wasn't sure whether or not I was imagining it, but there seemed to be a dim glow seeping out from under it.

I began to walk around the perimeter to see if there was another entrance or perhaps some kind of gap in the fence. I had reached the halfway point and had found nothing at all when I began to wonder if I was going to need some more specialist tools to break into this place. Some kind of angle grinder, perhaps.

But then, just around the next corner, I looked up and saw that there was a break in the razor wire. Bits of it were dangling free on either side. If I could just get up to the top of the fencing, I should be able to find my way over without cutting myself to ribbons. I grabbed hold of two of the metal posts in front of me and pulled myself upwards, searching for a foothold. The metal was quite smooth, but here and there were bolts holding it all together and I could just about get them to take my weight while I searched for handholds higher up.

After a few more minutes of struggling I found myself dangling from my hands at the top. All I had to do now was to pull my legs up somehow, swing them over and make my way down the other side without breaking anything important in the process. This turned out to be somewhat less straightforward than it initially appeared, particularly as the top of the fence posts were a little on the sharp side. However, I did eventually succeed in getting myself over them and, in a short while, I was down on the ground again, inside the fence.

I adjusted my scarf once more so that only my eyes were visible and set off back towards where I'd seen the fire exit. When I got close, however, I could see that it was already open and that someone – a man – was standing outside it, smoking a cigarette. I ducked down behind a nearby bin, watched and waited for them to finish. After what seemed like half an hour, he took his last drag, dropped the butt on the ground and ground it beneath his heel. He took a deep breath of the evening air, then turned and went back into the building... leaving the door propped open with a fire extinguisher.

I gave him a few minutes to clear the area and then poked my head inside the door. On either side of me there were a whole series of racks full of computer equipment. They were what was giving off that soft electronic glow that I'd noticed under the door earlier. The rest of the area was in darkness but I was acutely aware that, if anyone did suddenly decide to switch on the lights, I was currently very exposed. So I lowered myself to the floor and scuttled, crab-like, towards the rear of the cabinet nearest me on my right. Once I'd manoeuvred myself out of sight, I paused for a moment to allow my pulse to drop to something a bit closer to normal.

At this point it struck me that this seemed a very peculiar place to be keeping a pair of alpacas, assuming that Dolores and Steven were indeed still alive and in good health. Perhaps I should have tried the Penzance or Cumbernauld sites. Perhaps I should have tried somewhere else altogether. Perhaps I should have given up on the alpacas and stayed at home.

While I was thinking this through, I heard footsteps coming in my direction so I shrank still further into my hiding place until I came up against something soft. I reached out with my hand to find out what it was and came into contact with a human thigh.

'What the——?' I began, but I was cut short by a gloved hand enveloping my mouth. At the same time, something sharp and pointed was pressed into my back. I froze, barely daring to breathe. My pulse was back to where it had been a few minutes previously, and some.

'Mmffth,' I said.

'Don't say a word,' said a female voice in my ear.

Chapter 4

A few feet away from me, I could hear someone – presumably the guy who had been smoking – tapping away at a keyboard. Something beeped and he pressed a few more keys. It beeped again and then I heard the sound of a fan kicking in as a computer started up. This was clearly what was supposed to happen, because I soon heard the footsteps again, this time going away from me. Under normal circumstances, I would have relaxed at this point, but I still had the business end of a knife sticking into my back.

'Mmffth,' I said. I also still had someone's hand over my mouth.

'Not yet,' said the female voice in my ear. We stayed in position for what seemed like half the night, although it was almost certainly not much more than five minutes. Then my captor relaxed her grip.

'OK,' she said. 'I think we're safe for now.'

'Well, that's good,' I said. 'Would you mind taking that knife away? Just in case you do something by accident?'

'If I did do something, it certainly wouldn't be by accident. But then again, there's not a lot I can do with my front door key,' she added, removing the source of the pain in my back. She reached around and waggled a bunch of keys in front of my face. 'Had you fooled for a moment, though.'

I turned my head to look round at her. This didn't actually reveal much, as she was dressed in black motorbike leathers, with a black balaclava covering her face. The voice, despite being muffled, gave a strong impression of having had the benefit of an expensive education, during which it had almost certainly found itself in great demand on the hockey pitch.

'So you're not actually armed?' I said.

'Good god, no. You go into a situation armed with something deadly and all you actually end up doing most of the time is give permission to someone more competent than you to kill you. I'm pretty bloody useless at that sort of thing, so chances are if I were to start brandishing a bloody great knife around, whoever I was attacking would grab it off me in no time and use it against me.'

'So, despite dressing like one, you're not a ninja, then?'

'As if. Probably shouldn't be telling you this, come to think of it.'

'It's OK, I'm safe.'

'Well, you say that. But I wonder if I should be taking anything you say at face value. After all, you *have* broken into this place.'

'Wasn't me who cut the razor wire, though, was it?'

'Fair point. Hmmm. Talking too much. Bit nervous.'

'It's understandable.'

'I don't do this kind of thing very often.'

I was going to say, 'Me neither,' but it wouldn't have been entirely true. Something about the way she was acting suggested to me that *she* wasn't telling the whole truth either.

'Exciting though, isn't it?' she said. 'Stay there a moment.' The woman scuttled forwards out of our hiding place, got to her feet and then shimmied her way through the racks of servers until I could just make out her shadowy presence at the doorway peering into the rest of the building. Then she came back and knelt down next to me. Despite her claims not to be a ninja, she gave a decent impression of one in the way she moved.

'OK, it looks all clear now,' she said. 'First question is: who are you? Second one is: are you on my side?'

'Tom,' I said. 'And I'd need to know a bit more about you before giving an accurate answer to your second question. But if it helps, I'm almost certainly not on theirs.' I pointed towards the doorway.

'That'll have to be good enough for now, I guess. Come on.' She pulled me to the front of the rack and stood up. I cautiously picked myself up as well.

'One thing, though,' I said. 'Who are you?'

'Ada,' she said, holding out a hand.

I gave her hand a tentative shake.

'Hi, Ada,' I said. 'Are you on *my* side, though?'

'I guess we'll find out whose side we're both on soon enough. Follow me.'

Ada made her way towards a door. I hesitated for a moment, wondering what I was getting myself into. However, she seemed to have slightly more idea of what she was doing than I had, so I decided to do as she suggested and follow her – at least for now.

The temperature outside the server room was no warmer than inside. The corridor leading away to our right was in darkness, but the one on our left had patches of light here and there which I guessed must come from offices. There were unlabelled cardboard boxes randomly piled up on both sides with the occasional one or two stacked on a trolley waiting to be taken on an outing somewhere. Every now and then the boxes were punctuated by a line of fridges, standing to attention against the wall as if in an identity parade.

We headed off to our left and crouched down behind the first heap of boxes.

'Can you see anything?' I whispered.

Ada peered out. 'We're clear to the next lot on the opposite side,' she said.

'OK.'

She leapt up and dashed across the corridor to a position behind the next pile of boxes. I followed after her, just in time to avoid being noticed by someone emerging from an office ten yards or so further on. He wandered up the corridor away from us, singing as he went, in a voice that I found very hard to believe matched whatever was coming out of his earbuds.

Ada turned to me. 'Fancy a nose in that guy's office?'

'Yeah,' I said. 'Why not?'

Ada stood up and scurried away towards where the man had emerged from. I followed her, and soon we were both inside the office. The only furniture in there consisted of a desk and a chair, on one side of the room. The far wall was lined with more cardboard boxes that were similar to the ones taking up space in the corridor outside. Sitting on the desk next to a computer were a high-end microwave, a coffee machine and a bread maker. Behind the desk, someone had taped a joke notice that read 'Number of days since last critical incident: 0'.

'What do you reckon's in these boxes?' I said. 'Also, why the white goods on the desk?'

'Maybe he gets hungry from time to time,' said Ada. 'But that's not important right now.' She sat down and peered at the computer that was sitting on the desk. 'Well,' she said. 'Whatever they've got here, they're still working on it.'

I went round behind her and looked at the screen. It was covered in what I recognised, from my brief time working with Dorothy, as some kind of programming language.

'What does it do?' I said.

'Not a clue.' She took out her phone and took a picture of the screen. 'I'll get someone to check that out later. OK, let's have a gander at his emails.' She clicked on the mail icon and started scrolling through. 'Well, that's interesting,' she said, pointing to the third one down. It was asking if the programmer would have the 2.5.1 version ready in time for the convention next week.

'Why's it interesting?'

'Look who it's from.'

It was from Robert J Merritt III. 'Oh, him,' I said.

'What do you know about Robert Merritt?' said Ada, narrowing her eyes.

'Oh, nothing,' I said. 'Think I read about him some time. Rich American bloke.'

'He is. Very rich indeed. Big in all sorts of things, but especially gambling.'

'So what's he doing sending emails to a low-level minion developer over here?'

'That's the interesting question.'

'You don't sound very surprised,' I said.

'Let's say I had my suspicions.'

'Hold on,' I said. 'I think I can hear footsteps.'

'Shit. Get behind the door.'

We both rushed over to the opposite side of the room, concealing ourselves just in time before the programmer returned. I peered round the edge of the door to see him sitting down at his desk and staring at the computer. Then he stood up and started looking around, with a frown on his face. I quickly withdrew my head and stared at Ada in panic. She acknowledged this by pointing with her eyebrows towards the rows of boxes. I nodded back, realising what she had in mind.

I reached out and gave the shelving unit a tentative wobble. The top teetered back and forth a couple of times and then, one by one, the boxes started to topple over, emptying hundreds of rubber ducks with the face of Tony Blair over the office's occupant.

'Come on!' said Ada and I followed her out of the room into the corridor.

'Left, or right?' I said.

'Right,' she said. 'Come in by the fire exit, leave by the front. Grab a few of those as we go.' She pointed to the boxes on either side of us. I got hold of a couple of them and dumped them on a trolley. Then I ran after her, pushing the trolley in front of me. Every so often, I deliberately rammed a particularly promising pile of boxes with the intention of knocking it over. Anthropomorphic rubber ducks spilled out everywhere – Posh Spice, Simon Cowell and Nigel Farage all added to the mix cluttering up the floor behind me.

We weren't quite halfway to the end of the corridor when we heard multiple footsteps behind us and voices shouting at us to stop.

'Ignore them and keep running,' said Ada. 'We're nearly there!'

'What are we going to do when we get to the end?' I said.

'We'll think of something!'

'What though?'

'Don't know. Keep going.'

Behind us, there was a thump and a cry of 'Shit!' as one of our pursuers slipped on a duck and fell over.

'One down!' I said.

Then I heard the first shot. The bullet zinged over our heads and buried itself in the wall a little way further down the corridor. I heard another cry of 'Stop!' and I turned around to see a man dressed in camouflage gear taking aim in our direction. I glanced across at Ada.

'What do we do?' I said. 'They're shooting at us!'

'This.' She grabbed the trolley from me, spun it round, and launched it backwards towards the man. It was gloriously accurate and it caught him full on, knocking him right over and causing him to loose off a couple of rounds into the ceiling above.

By now, we had reached the end of the corridor. We launched ourselves against the front door of the factory, causing it to crash open. Ahead of us was the main gate, where the chain was still holding it firmly in place.

'Now what?' I said.

Ada paused, staring at the padlock. 'Any good with locks?' she said.

'If I was, I'd have come in that way in the first place.'

'Fair point. OK, left, then. No, right. No, left!'

She took off along the perimeter of the building, with me trailing in her wake. I was developing a horrible feeling that this Ada, whoever she was, had about as much of a clue as to what she was doing as I had, and this wasn't a comforting thought.

'What's the plan?' I shouted after her. 'You know there's a man with a gun after us?'

'Of course I do,' she said. 'That's why I'm running away.'

'Maybe if we just surrender?'

'If you want to do that, I'm not stopping you. Oh shit,' she suddenly said. I stared into the darkness ahead and realised that there was now someone heading towards us in the opposite direction. We were caught in a pincer trap.

'Back in here!' she cried as we drew level with the fire exit. As soon as I was in, she removed the fire extinguisher that was holding the door open and pulled it firmly shut. We both paused for breath before heading back towards the interior of the building.

'I hate to sound a repetitive note,' I said, 'but now what?'

'I think it's about time you came up with a suggestion, isn't it?' said Ada. 'I'm doing all the heavy lifting here.'

'Where do they keep the guns, I wonder?'

'Know how to fire one safely?'

'I'm sure I can work it out.'

'And *I'm* pretty sure you can't. So let's put that idea to one side. OK, if we go left, we end up where we started, with the difference that the floor is now covered in rubber ducks.'

'Turn right, then?'

'At last! A sensible suggestion!' She paused. 'But before we go, let's cause a little mayhem.' There was a big red button on the wall by the door. She pressed it with the palm of her hand and immediately every single piece of equipment in the room, along with the lights, shut down, to be replaced by the soft red glow of emergency lighting.

'Nice.' I waited by the door.

'After you, then,' she said, giving me a nudge forward. I hesitated for a moment and then there was a loud thumping on the outside of the fire door. I lurched off into the red-tinged gloom to my right.

I hadn't gone more than a couple of paces when I almost tripped up over a step. I dropped down to my knees and felt the ground with my hands.

'Does your phone not have a torch or something?' said Ada, coming along side me and shining a light into the darkness ahead.

'I wanted to preserve the battery,' I said.

The passage ahead of us led up a staircase which turned around a corner halfway up.

'What do you reckon?' said Ada. 'Fancy the roof?'

'If we want to get trapped,' I said, 'it sounds a great idea.'

At that point, there was a crash of a door hitting the wall and sounds of shouting from the front door to the unit, back at the opposite end of the corridor.

'I think it might be our only option,' said Ada, running off up the stairs, waving her torch around as she did so. I followed after her. At the top of the stairs, there was a fixed metal ladder leading up to a skylight. Ada was already at the top of this, fiddling with the catch to open it. Eventually, it swung open and we both crawled out onto the roof.

'OK,' said Ada, closing the skylight again. 'First thing we do is find something heavy to weigh this down.' Luckily, there were a couple of loose breeze blocks lying around nearby and we picked them up and put them in place.

'Is that enough?' I said.

'It'll have to do. We need to find another way down now.'

Ada ran over to the edge and knelt down, peering over. As I reached her, she pointed towards a yellow skip that had been placed on the other side of the metal perimeter fence. It appeared to be filled with discarded cardboard boxes.

'Bet you they're full of rubber ducks,' she said.

'You're not proposing to jump into that, are you?'

'Not exactly.'

'Good, because the top of the fence is a bit sharp even without the razor wire and I'm not in the mood for impaling myself today.'

'Give me a hand here,' said Ada, reaching down and grabbing the top of a nearby downpipe.

'Sorry, what?'

'Just don't ask questions and give me a hand. Look, it's loose.'

She was right. The downpipe wasn't properly attached to the wall and with a bit of effort, we managed to prise it free.

'Now,' she said. 'We lift it up so that it rests on the top of the wall.'

'What?'

'Just do it.'

'But—'

'Please?'

'But why—?'

'Have you got a better plan?'

'Well, no. I guess I was just questioning my position in the command and control hierarchy here.'

Ada just stared at me. Or at least, I think it was a stare, because it was hard to discern the nuances of her expression behind her balaclava. Whatever it was, there seemed to be a considerable level of intent behind it.

'Oh, all right then,' I said eventually.

I did as she asked and, after a certain amount of heaving and straining, we ended up with the drainpipe balanced on the top of the metal fence with the end dangling over the skip.

'Right,' she said. 'Hold this end as firmly as you can.'

'I still don't see what you're...'

Ada was placing an experimental foot on the sloping downpipe.

'Oh my god, you're not really going to—?'

'Yup,' she said. 'Grip feels good.' She took a deep breath. 'Off we go!' And, without a backward glance, Ada ran down the length of the pipe, over the fence and into the skip.

I stared down at her, terrified. She waved back at me, urging me to do the same. She had reached up and was holding on to the bottom end of the pipe to steady it. She beckoned to me again, with more urgency this time. I still hesitated. For one thing, I really wasn't sure if the tread on my shoes was sufficient to grip on a narrow tube. For another, my sense of balance wasn't great at the best of times and this was pretty close to the worst of times.

My indecision lasted another couple of seconds and then I heard the sound of someone hammering at the skylight behind me. I was out of choices. I stepped out onto the pipe and began my journey down. I instinctively knew that the best approach was to do exactly what Ada had done, but sheer terror overrode instinct and I opted for a more cautious approach. Halfway across the gap between the roof and the fence, the folly of this became apparent as I slipped over sideways. As I went down, I grabbed hold of the pipe with my left hand and somehow managed to hang on. After a brief pause to catch my breath, I brought my right hand up and succeeded in swinging my legs up so that I was now suspended, sloth-like from the mid-point.

I inched my way down from there towards the fence itself, anxiously wondering how I was going to get over the fence without having my lower limbs shredded into a bloody pulp. But by the time I got there, Ada had found a discarded dust sheet in the skip and had draped it over the razor wire, so I managed to wriggle my way over it and into the skip, where my fall was cushioned by half a dozen boxfuls of rubber ducks with the face of a distinctly racist caricature of Barack Obama.

'Thanks,' I said.

'You're welcome,' said Ada. 'Now let's get out of here, before... ah.'

She was looking up at the far end of the drainpipe. I turned and saw what had caused her to pause. A man with a gun was staring down at us. Holding the gun in both hands, he slowly raised it to chest height and aimed it carefully in my direction. Then he doubled over at the waist before toppling over backwards as a sharp thrust from Ada with the drainpipe caught him in the groin.

'Ouch,' I said. 'Nice work.'

'Well, I was holding the thing anyway,' said Ada. 'Come on, let's get moving.'

We scrambled out of the skip and I followed her as she ran over to the car parking area where a gleaming red and silver motorbike was propped up. I watched as she unclipped a helmet, put it on and swung her leg over. Then she turned the key and the engine roared into life.

'Well, come on,' she said.

'I haven't got a helmet,' I said.

'So what do you plan on doing? Call a taxi? Our friend with the gun may not wait that long.'

I dithered for another few seconds, then climbed on behind her.

'Should warn you, I go quite fast, so hang on tight. Don't be shy. Where do you need to get to?' Before I could answer we were moving.

She wasn't kidding about going fast. I swear the machine went from zero to sixty in a couple of seconds before it had even left the car park. I threw my arms around Ada's waist and clung on as if my life depended on it. Which was the right thing to do in the circumstances, because right now it almost certainly did.

By the time we got back to the caravan, an hour or so later, I was a gibbering wreck. Still, I'd somehow survived the evening's events, so there was that.

'Nice place you've got here,' said Ada, raising an eyebrow as she admired the exterior of my father's domain. From her position on the step, μ gave her the best hard stare she could muster. Despite her disappointment at being brought here to live, she was still fiercely protective of her territory and would not brook the slightest criticism of it, either explicit or implicit.

'It's only a temporary arrangement,' I said, releasing my grip on Ada and swinging my trembling leg over the bike. I hopped about for a moment, re-establishing my connection to planet Earth. It would be some time before my legs believed in the existence of solid ground again.

'Thanks for the lift,' I said. 'And for everything else.' I paused for a moment, wondering how to keep the conversation going. 'Look, who are you working for and what happens next?' I said.

'What happens next,' said Ada, 'is that you go into your snug little caravan, get a good night's sleep and forget tonight ever happened.'

'Not going to happen,' I said. 'And you didn't answer the first question.'

'I didn't, did I?' she said. Then she kicked the accelerator and roared off into the night.

Chapter 5

I didn't sleep well that night. There were so many questions fizzing around in my brain that it was difficult, at times, to even try to focus on just one to lie awake wondering about. Who on earth was Ada, for one thing? Also, what the hell was going on inside the Luxxy Duxxy factory? What did it have to do with the theft of Margot Evercreech's alpacas? And why was it so important that whoever was doing it felt obliged to protect themselves with armed guards? I had perhaps grown a bit too used to people trying to kill me in recent years, so the significance of the last one of these didn't strike home at first.

The involvement of Robert J Merritt III was especially troubling. Was this really something I wanted to get any more involved with? Perhaps it was time to stop having daft adventures and to settle down and stay safe long enough to provide Ali and Patrice with sufficient of the necessary to help them have a family.

So for the next few days, I went back to my old routine of complaining about the spoiled milk, cooking indifferent spaghetti bolognaise for the two of us and drinking too much cheap alcohol in the evenings. It was an unambitious life but it was not entirely without value, I felt. For one thing, I am sure that my father's life was considerably improved by my presence, if only because – so far at any rate – I had managed to stop him from investing in any more daft cryptocurrency schemes. Even Wally the dog appreciated my presence – although that might simply have been because of the additional diversity of the smelt environment resulting from the presence of two male crotches rather than just one.

Two days into this new low-ambition life, I had a call from the fertility clinic to make an appointment for a blood test and general health check prior to taking up my role as donor for Ali and Patrice, plus another one for a few days later to do the actual procedure, on the assumption that the tests didn't show up anything peculiar. Somehow this added to my general feeling that I was now taking life more seriously. I had responsibilities.

The first check-up passed off without incident, but on the way back all my well-laid plans began to fall apart. An earlier passenger on my train had left a copy of one of the tabloid newspapers on the seat next to mine. I didn't have anything else to occupy me on my way home so I started to read a feature on the rise of the online gambling industry, and that's how I found out about the very first *CyberGambleCon*, an internet gambling convention, to be held at the Robert J Merritt III Convention Center in Las Vegas the following week. While I was skimming through the piece and wondering what the chances were that my father had already been sucked into any of this stuff, a call came though on my phone. The number looked familiar.

'Hello?' I said.

'Ah, Winscombe,' said a female voice. 'Evercreech here.'

My heart sank.

'Hello, Margot,' I said. 'How are things? How's the archiving going?'

'Never mind that,' she said. 'How are my alpacas?'

'They're... fine,' I said. I wasn't exactly lying because there was every chance that they were indeed fine. Wherever they happened to be.

'Good,' said Margot. 'Because I think it's about time they came home to live with mummy. I'm sure they must be missing me.'

'Oh they are,' I said.

'So then. When can you bring them over?'

'Let me check,' I said. 'Oh, hold on, I'm going into a tunnel.' I held my hand over my mouth and shouted something loud and incomprehensible at her, then cut the call. It probably wasn't terribly convincing, but there was an outside chance she'd fall for it. Then I switched my phone to airplane mode while I pondered what the hell I was going to do.

I couldn't face the prospect of telling Margot that I'd lost the alpacas. I had to get them back. But the only clue I had was a vague idea that the Merritt organisation was somehow involved, and given that they – or at least the Basingstoke branch – seemed to be the kind of people who shot at intruders, I felt that it was unlikely that they would respond helpfully to my potential enquiry.

I looked down at the paper on my lap. Then I had a wild idea.

All roads seemed to be heading to Las Vegas. What if I went to the convention? After all, chances were I'd be able to find someone there from the Merritt organisation – maybe even Robert J Merritt III himself – who I could safely collar to ask them about what was going on with Dolores and Steven. I could be back in less than a week if everything went to plan. And, if I got myself a flight for the very next day, I would be nicely sorted out from jet lag by the time the convention opened. My cash resources were still at an all-time high, thanks to having spent almost nothing for the last six months, so paying for my ticket wasn't going to be a problem.

It was, I had to admit, an absolutely brilliant plan.

I called Margot back.

'Sorry about that,' I said. 'Bit busy right now. How about the week after next?'

I spent the rest of the journey home trying to find the cheapest return flight to Las Vegas. I also reserved a place at the *CyberGambleCon* convention, just in case it sold out before I got there. Finding somewhere to stay was more problematic as everywhere seemed to be pretty much booked up and the only alternatives were to pay for somewhere stupidly expensive or to slum it in a place going by the moniker of The Heartbreak Motel. I decided to save my cash and go with the latter.

The Heartbreak Motel was a couple of miles away from the main Vegas Strip, sandwiched between Arlene's Liquor Store and a branch of Red Lobster. The place, like its name, had a half-baked Elvis theme to it. The restrooms in the lobby were labelled 'Hound Dogs' and 'Hard Headed Women', while a notice on the counter made the somewhat defensive plea 'Don't be Cruel' to any potential clientele. Next to this, a selection of business cards was stacked up in a holder discreetly labelled 'Are You Lonesome Tonight?' A faded poster for the film of *Blue Hawaii* clung on to the wall behind the counter by its finger-nails, with the top left-hand corner dangling free. The female receptionist was in full late-period Elvis attire, complete with resplendent fake sideburns, although her vocal impression only extended to the phrase 'Uh-huh', which she made full use of as her response to every single enquiry.

'Hi,' I said. 'I have a reservation.'

'Uh-huh?'

'Yes. Name of Winscombe.'

'Uh-huh.'

She produced a credit card scanner. I fished out my wallet and withdrew my card and inserted it into the machine.

'Uh-huh,' she said, gesturing for me to enter my PIN. I did so. She took the machine away from me, removed the card and handed it back to me.

'Thank you,' I said.

'Uh-huh.'

She fetched a key and pushed it in my direction. It was attached to a piece of plastic which bore the ominous moniker 'Shake, Rattle and Roll'. I turned round and nodded in the direction of where I assumed the rooms were.

'Uh-huh?' I said.

'Uh-huh,' she confirmed. I felt quite pleased that I was now fully fluent in Elvis.

On the way to my room I bumped into a number of similarly dressed folk and for a brief moment I wondered if they were all staff as well. However, a flyer shoved under my door advertising the '45th Annual Presleyana Convention' suggested an alternative explanation. It was entirely possible that I was, in fact, the only civilian staying there.

The carpet leading to my room appeared to store sufficient static electricity to power a small car, but as it inconveniently discharged itself into my hand as I grasped hold of my door handle, I was unable to verify this. The room itself was relatively clean, although as I unpacked it became apparent that I was sharing it with a number of sitting tenants in the shape of a family of cockroaches. However, I was so exhausted when I finally crawled into bed that neither their incessant chattering nor the television blaring through the wall that divided me from the room on one side nor the heavily armed police raid that seemed to be going on in the room on the other side managed to keep me awake for more than a couple of seconds. Even the brief off-key rehearsal of 'Crying in the Chapel' from somewhere down the corridor at around four a.m. only disturbed me for a minute or two.

The next morning I woke up feeling thoroughly refreshed. I washed and dressed and thought about the day ahead. It was Sunday and the convention didn't start for a couple of days, which gave me a good opportunity to orient myself and work out my plan of action for *CyberGambleCon*. But first I needed breakfast. I left my room and headed for the front desk to find out where to go. However, I was clearly too early for the team as there was no one else in evidence when I got to reception, so I stepped out of the front of the motel, took a deep breath of the warm desert air and picked a direction to walk in.

I breakfasted on an Egg McMuffin from the McDonald's on the opposite side of the Red Lobster, followed by a Boston Kreme doughnut from Dunkin' Donuts a little further down the same block. In my defence, my stomach was giving me mixed signals and letting me know that it still quite hadn't worked out what the correct time of day was. And in any case, I could be about to embark on something dangerous, and I deserved to indulge myself if there was a non-zero chance that this might turn out to be my last meal.

I decided to work off my breakfast by walking into the centre of the city. Las Vegas on a Sunday morning had a hungover air to it. It was as if the whole city had woken up in an unfamiliar bed with only the haziest recollection of what it had got up to the previous night and, having endured a stilted conversation over breakfast with a complete stranger, was now setting off back home still wearing its party clothes, wondering if a visit to confession might be on the cards later in the day.

The Robert J Merritt III Convention Center turned out to be a vast cavern of a place, just across the road from the El Gran Pelícano hotel and casino. The front doors were all barred, so I wandered round to the back where there was a steady stream of vans going in and out of an underground parking lot. The ones on the way out were emblazoned with folksy airbrushed logos depicting garden ornaments, while the ones heading in promised a glorious future filled with untrammelled riches if you placed your life savings on just one sure-fire bet. For a moment, I imagined myself spending a few days in the company of garden gnomes rather than internet gamblers and the world seemed a far more wholesome place.

No one seemed to be checking any details of arrivals at the tradesmen's entrance and so I wandered down the ramp into the parking lot beneath the hall, where groups of men and women wearing matching T-shirts were mustering around each of the parked vehicles. Others wearing prominent Bluetooth earpieces and carrying iPads were fussing around attempting to herd them into some sort of order, while complaining that they were struggling to get a signal underground. I carried on walking until I found the stairs going up to the ground floor.

As I reached the top, one of the iPad women collared me and I instantly expected to be thrown out. She had long blonde hair and perfect teeth that glinted alarmingly whenever the light happened to catch them.

'Hey!' she said, 'Where have you been and where's your uniform?'

I looked around, but there was no one behind me.

'Me?' I said.

'Oh, never mind, just get your ass over to Team C.' She waved me away contemptuously towards a motley selection of half a dozen workers loitering nearby. I made my way over.

'Hi,' I said.

The others grunted, apart from one guy with a moustache that looked as if it required feeding twice a day. He took a long draw on his cigarette, then gave me an evil stare and muttered, '*No eres Diego.*'

'I'm sorry?' I said.

'You are not Diego.'

'Ah, I see,' I said. 'Right. No, I'm not Diego. Diego is indisposed.'

My new friend took one last drag before grinding the butt underneath his heel, before making a sort of 'Pfft' noise with his lips.

'Pleased to meet you,' I said and he sneered back at me.

'OK, listen up!' shouted the woman with the iPad. 'Team A to concession 256 on the ground floor with the main booth from store room 2. Team B to the primary AV suite with the FX kit from store room 1— I'm sorry? What? Just ask, Gabriela. Well, someone's gonna know. Well, all I know, Roberto, is it's on the ground floor. Just fucking ask, man. Use your initiative. What? Oh, you most certainly are being paid to use your initiative, Roberto. Jesus.' She paused for breath and shook her head in disgust. 'OK, Team C to the Marshall Suite on the mezzanine with the private pitch pod from store room 3C. And take special care with the fish tank, guys. No, Carlos, we've been through this. You're in Team B. I don't care if Josefina's in Team A – like I said, you're in Team B. No, I don't do transfers.' She gestured towards her iPad. 'This piece of shit software barely lets me read the allocations I've already made. Any questions?'

There was a general mumbling from the assembled parties, but no real dissent emerged.

'Good,' she said. 'OK, one last thing – need a word with Diego.'

No one moved.

'Diego!'

It took me a full ten seconds to realise this was directed at me.

'I'm… not Diego,' I said.

'You are now, kid. I couldn't change your name if I wanted to.'

I went over to her.

'OK, Diego,' she said. 'Here you go.' She handed me a T-shirt. It had a logo featuring a pelican holding a sash on which was emblazoned 'Robert J Merritt III Convention Center'. Underneath this, the words 'Crew Member' had been appended. I took my top off and substituted my new outfit. I was now a fully paid-up member of the gang.

'Do I get paid in cash or gambling chips?' I said.

'You actually *want* to get paid in chips?' she said, raising an eyebrow.

'Ha, no.'

I went back to my group, who were already unloading stuff from the van onto a large trolley. I pitched in and soon I was accepted as a fellow member of Team C – by everyone, that is, except evil moustache man.

Once we'd got the first load onto the trolley, we headed off to the service lift and found our way up to the mezzanine, where we located the required place in the Marshall Suite at which to erect the private pitch pod. However, we were unable to get on with it because the previous occupants were still busy dismantling an immense fibreglass unicorn. I wandered over to the balustrade overlooking the main hall, where María, a short, muscular woman of indeterminate age with spiky hair, was taking a swig from a can of Pepsi Max.

'I wonder if they'd let me have that thing cheap,' I said, gesturing towards the unicorn behind us.

'Better spend your money on that than these mythical beasts,' she said waving her hand towards the array of booths that were springing up below us. She enunciated each word carefully, and there was a slight Latin-American twang to her voice.

'You don't gamble, then?' I said.

'I run a strictly cash-based economy,' she said. 'Much safer that way. So where are you from, Mr Not-Diego?'

'England,' I said.

'You don't have to keep up that cute accent, you know.'

'Hey, it's real. Listen,' I said. 'What's going on with these guys? How did you get recruited?'

'Same as everyone else here. From the border labour exchange.'

'We are the ones they call illegals,' added one of her colleagues, Jesús, joining us.

55

'I prefer the word "paperless",' said María, shaking her Pepsi to make sure it was empty. 'It sounds kind of modern and it's much less judgemental.'

'Fair enough,' I said. 'So what's the big boss like?'

'Little Miss iPad?' said Jesús.

'No, I mean the real big boss.'

'You want to know about God?' said María, amplifying this by extending a horizontal hand as high as she could reach upwards.

'Maybe not quite that far,' I said, placing my own hand at neck height. 'I was actually asking about Robert J Merritt III.'

A brief expression of fear flashed across Jesús's face. 'What do you want to know about Merritt?' he said.

'Well,' I said. 'What I really want to know is why he might have arranged to have my alpacas stolen?'

The pair of them looked at me as if I was insane.

'OK,' I said. 'Maybe that does sound a little odd. You see—' I was interrupted by a loud crash behind me. I turned round to see the large fibreglass unicorn lying in several pieces on the ground, with a group of people in overalls staring down at it and scratching their heads.

'Well, I guess we're clear to move in,' said María. So I never did get round to explaining the story of the alpacas.

I spent my first day as a crew member at the Robert J Merritt III Convention Center assembling and applying a first coat of paint to their private pitch pod, which seemed to be a space for schmoozing high rollers to the extent that they grew hazy about how much of their hard-earned cash they'd agreed to stake on Merritt's dubious games of chance. I tried to find out as much about Robert J Merritt III as I could, but the others on my team turned out to be more than a little cagey on the subject. The deal seemed to be that, if you wanted to keep your job and not get shopped to the immigration authorities, it was best to avoid asking too many questions.

The first thing I did find out about Merritt was that he seemed to be a big collector of exotic animals, although, judging from the picture of him surrounded by big cats that adorned the side of one of the booths, the fauna that he was really keen on tended to be, both figuratively and literally, somewhat higher up the food chain than mere alpacas. The second thing I found out was that he was, as I had hoped, due in town this very week and would be staying in the presidential suite at El Gran Pelícano just across the road. I wondered if he might be dropping in the next day to inspect our work, just to be sure that we'd kitted everything out to his satisfaction. But on reflection, it was unlikely that one of the richest men in the world would be interested in checking that we'd drilled the holes in his MDF boards the regulation 2 ¾" from the edges, so if I wanted to make contact with him, I had to think of some other way.

When I emerged from my day's work, dusk was beginning to fall in downtown Las Vegas, although the ubiquitous neon more than compensated for the lack of light from any celestial object. A young woman dressed in an impossibly bright blue satin jumpsuit came over to me and enthusiastically thrust a handful of flyers for the convention at me.

'It's all right, I've already got my ticket,' I said.

'AWESOME!' she replied, as if I'd just won the lottery and signed my entire winnings over to her on the spot. 'Y'all have a great day!'

'I most certainly will,' I began, but she'd already moved on to her next mark. I wandered over to El Gran Pelícano and peered inside the front door of the casino. The slot machines were already getting some heavy action from low-stakes punters and it struck me that so much of gambling was light years away from the James Bond glamour of dinner jackets and beautiful women. Most of it was just desperate people trying to dig their way out of whatever hole they'd found themselves in. I resolved there and then never to allow myself to get involved in anything that required me to bet on anything. As I made this firm resolution, a little voice in the back of my mind was wondering how long it would be until I broke it, and it turned out to be quite a lot sooner than either myself or the little voice would have predicted.

Chapter 6

I trudged back to the Heartbreak Motel, which had somehow moved a full ten additional miles out of town since this morning. My brain was still stranded somewhere in a mid-Atlantic time zone and ready for its nightcap. Halfway home, I stopped off for a pizza that temporarily revived me, but by the time I rolled into the reception area, I was barely capable of grunting any kind of greeting to the woman behind the desk, which was fair enough because she wasn't that bothered about acknowledging my presence either. I staggered back to my room, forgetting once again to earth myself before I grasped the door handle, although the resulting shock did at least have the temporary effect of waking me up sufficiently to remember to brush my teeth before collapsing into oblivion.

Next day, I turned up for work at the Robert J Merritt III Convention Center again and once again, Diego was still missing in action, so I had a gig for the day. Or at least I had a gig until midday, when iPad Lady came up to our team as we were about to manoeuvre a fish tank into a pod.

'OK, you guys,' she said. 'I have a new mission for one of you, should you choose to accept it. Mr Merritt's handyman has called in sick and we need someone to carry out some rewiring in his suite at the Gran Pelícano.'

No one volunteered. I tentatively put up a hand. Setting aside all other considerations – and there were several – this would be an excellent way to get into the penthouse and find out more about what Robert J Merritt III was up to.

'OK, Diego,' she said. 'Looks as if the gig is yours. You do understand electrics, right?'

'Um… yes, of course I do,' I said. Of course I knew everything about electrics.

'You sure?' she said.

'Definitely. One thing though – why can't he get someone from the hotel to do it for him?' I said.

'Mr Merritt owns the Gran Pelícano.'

'Ah,' I said. 'So I'm being seconded to the hotel staff, right?'

'You're getting a day's work as a stand-in handyman. That's it.'

Needless to say, handymanship was an area in which I possessed nothing which in any way approached the concept of a tangible skill. Now if it simply involved bashing something with a hammer or maybe tightening a nut or two, I could almost certainly bluff my way through it. However, I knew absolutely nothing about electrical wiring beyond very occasionally having to re-wire a plug on one of my father's more ancient appliances, and I had no idea about American wiring whatsoever. There was a distinctly non-zero chance that I could either electrocute myself or set fire to the entire building. Or, more likely, both. Still, at least this way I was definitely going to get a peek inside the big man's domain.

'Do I get any tools?' I said.

iPad Lady reached down and handed me a box that, judging by the weight of it, contained a sufficient number of gadgets to wire up a small factory.

'OK, no problem,' I said, trying not to gasp under the weight. 'When do I start?'

iPad Lady looked at her watch. 'Five minutes' time. So if I were you, I'd get my ass over there right now. Tell 'em Jolene sent ya.'

I bade farewell to my newfound companions in Team C, noticing in passing that none of them seemed that bothered that I'd been selected for promotion. It was almost as if they were relieved to have been left carting fish tanks around the place.

I walked out of the hall once more and headed over to El Gran Pelícano, taking a circuitous route in order to avoid the attentions of the bright blue satin jumpsuit women. I walked up to the front desk and enquired as to how I could get to Mr Merritt's suite.

'Mr Merritt ain't expecting no visitors,' came the flat reply. The man who gave it didn't even bother looking up at me and continued monitoring the bank of security cameras in front of him.

'But I've been ordered to come over here,' I said. 'I'm the new handyman.'

The man at the desk sighed as if to express his deep displeasure at my continued presence. 'Mr Merritt is very particular,' he said, looking up at me over the rim of his glasses.

'But Jolene sent me over.'

'Jolene?' The man leaned back, took his glasses off and stared thoughtfully at me, then jerked his finger towards a door behind him. 'Take the elevator on the left. Goes straight to the top. And make sure your shoes ain't got no shit on 'em.'

Without thinking, I picked up each foot in turn and examined the sole. Well, they looked clean to me. I strode forward to the lift and pressed the button. After a minute or so, there was a soft 'ping', the doors opened and I entered a different world.

The lift appeared to have been cast out of a single block of finest Carrara marble, which was clearly impossible. But if it was veneer, I wasn't able to find any joins. The fittings were pure gold, apart from the Up and Down buttons, which had a pearl inset. Not that I actually needed to press any buttons, as a lift attendant in a finely tailored uniform was standing ready to do the honours.

'Up, please,' I said, realising, as I did so, that it was a somewhat superfluous request. The attendant bowed slightly and then pressed the button with a white-gloved hand. The doors closed and for a moment it seemed as if nothing more was going to happen. Then my stomach informed me that we were actually moving upwards at very high speed, like supersonic silk.

Thirty floors up, the lift slowed to a halt and there was another soft 'ping' as the doors opened. The attendant bowed and motioned for me to make my exit. In my crew member T-shirt and jeans, and carrying a hefty toolbox, I felt more than a little out of place in the mad palace I'd just walked out into.

It was clear from the décor of El Gran Pelícano's presidential suite that Robert J Merritt III of Arkansas and Las Vegas had the same sense of aesthetic restraint as King Ludwig II of Bavaria. The amount of gold in evidence on the walls of the corridors that led off in all three directions from where I was standing was sufficient to move the commodities markets several points if he were to decide to move out and sell it off. And the sheer size of the chandeliers made me wonder about whether the ceilings had been sufficiently reinforced to take their weight. The floors were covered in enough skins of dead animals to have once formed a decent-sized zoo.

'Hello?' I said, as the doors of the lift hissed softly together behind me.

The first door on the right down the corridor I was facing opened and a large bear of a man wearing a white coat emerged. He was wearing a badge with the word 'Don' on it. I wasn't entirely sure whether this was an indication of his mafia status or if it was his first name. On reflection, I decided to go for the latter.

'Who the fuck are you?' said Don.

'Jolene sent me,' I said. 'Maintenance.'

'I don't like the look of you,' said the man.

To be fair, I didn't like the look of him much either, although I felt it wise to keep that to myself.

'Know about electrics?' he said.

'I know everything about electrics,' I said.

'I bet you do,' said the man. 'I bet you do.'

I could already tell this was going to go really, really well. I picked up my tools and walked slowly towards where Don was standing waiting for me.

'OK, um, Mr Replacement Maintenance Guy, what did you say your name was?' said Don.

'I didn't,' I said. 'It's Diego.'

'Diego?'

'Yeah. Diego,' I repeated. 'Rivera,' I added on a whim.

Don pulled his glasses down his nose and peered at me over the rims, eyebrows raised in full military salute.

'My mother was an artist,' I said, improvising wildly.

Don folded his arms and gave me a sceptical look. 'Funny accent you got there, Diego,' he said.

'We have artists in England too, you know.'

'Is that so, Diego?' said Don. 'Is that so?' He continued to observe me for what felt like the best part of another minute before shrugging as if to admit that he hadn't managed to gather all the evidence he needed to establish that I was a complete fraud quite yet, but that there was very little doubt in his mind that he very soon would. I suspected that, in this, he was one hundred per cent correct.

'Follow me, then, Diego,' he said, with unnecessary emphasis on my assumed name. He began to walk off down the corridor, so I went after him, trying to go through in my head everything I'd learnt about electricity at school. There was definitely something about watts, volts and amps, but how they all fitted together was for now still a mystery.

Don stopped at a door which had a large sign on it saying 'Danger of Death: Do Not Enter'. He unlocked it and we went in. The room was the size of a small broom cupboard, but it seemed to have enough electrical equipment in it to power a small village, as well as giving off enough heat to provide an authentically sweaty sauna experience.

'So where's the icy lake for me to plunge into once I've finished?' I said.

'I'm sorry?' said Don.

'Never mind.'

Don gave a little shake of the head and waved his hand vaguely at the rack straight ahead of us. 'Seems to be an issue with the second-tier spofflewack throughput,' he said. 'Causing the plinge flutes to overflume and mulch.' To be fair to Don, that probably wasn't precisely what he said, but that's pretty much what my brain managed to take in.

'Right,' I said. 'Plinge flutes. Overfluming.'

'What?'

'Sorry, I'm struggling with some of the US terminology.'

'I'll draw you a diagram.'

'No, it's fine,' I said, anxious to bring this conversation to a rapid close. 'Just leave me to it. I'm sure it'll be obvious once I start disgorging the macerating fleugenvalves.'

I decided to fixate on a grey rectangular unit that was round about my eye level with a couple of switches and half a dozen lights that flickered in an irritatingly random manner. I toyed with the idea of seeing what happened if I switched it off and on again, but there's every chance that something could go horribly wrong if I did, so I just continued gawping at the lights.

Don stared hard at me, clearly wondering what kind of an idiot Jolene had sent him. I noticed that he was now sweating profusely, so there was perhaps a decent chance that he was also quite keen to get out of our cramped cubby hole. However, for the moment he continued to linger, like a fart in a lift.

'Honestly, I can get on with this on my own,' I said.

'You might wanna start down there,' said Don, pointing to a box at floor level which had a red light that flashed at regular intervals.

'Ah, *that's* where it is,' I said, squatting down. 'Usually in the UK, they put this bad boy at the top of the rack.'

'Is that so?' said Don, who sounded as if he would have found it more plausible if I'd suddenly announced that I was an alien visitor who had just stopped off for a cheeky toilet break en route to Proxima Centauri. I ignored him and continued to study the flashing red light box, willing him to get some kind of distraction that would drag him away from me.

Unfortunately for me, this particular unit was even more enigmatic than the one I'd been looking at previously. There were no switches at all on the front and even when I fumbled round the back of it, there didn't seem to be anything that felt like it would turn the thing off and on again, which was pretty much the limit of my skills in this respect.

Thankfully, at the precise moment when I was running out of ways in which to do nothing, Don's phone went off.

'Yeah,' said Don. 'No, I'm in with Diego… the handyman… come to fix the STS throughput… no, they're in my office… gimme a moment…' He ended the call and looked down at me, trying to decide whether to chuck me out there and then or to leave it until he'd done whatever he had to do back at his office. 'OK, Diego,' he said eventually, 'gotta go. Lemme know if you need anything. I'll be in my office down the corridor. Don't go away either.'

'Sure thing, Don,' I said, relieved that he was finally going to leave me alone. He took out his handkerchief, removed his glasses and wiped his face before leaving the room, closing the door behind him. OK, I thought to myself. Now what? Well, there was obviously no question of actually attempting to carry out any repair work whatsoever, so while I was here I might as well try and find out as much as I could about Robert J Merritt III and then hop into the lift and jog back to the motel before anyone noticed that I'd gone.

Before I did that, I took a closer look at the contents of the room. Now that I studied the racks more closely, I began to realise that they were filled with very similar equipment to the ones I'd seen at Luxxy Duxxy back in Basingstoke. There was also something about the aesthetics that reminded me of the installation at Luxxy Duxxy. One of them had the legend BTOKE taped to it, which meant nothing to me whatsoever. Neither did the one underneath it, labelled SPORE. I took out my phone and grabbed a few snapshots for later analysis.

I opened the door and peered out into the corridor. There was no sign of Don at all. I set off in the direction leading away from the lift, trying not to make much noise while simultaneously attempting not to look remotely suspicious. When I reached the end of the corridor, I hesitated for a moment, wondering whether to try left or right first, before opting for left.

It turned out that the left-hand corridor quickly terminated in a blank wall with a single door in the middle of it. There was a notice in the middle of the door instructing all potential visitors to KEEP OUT. I ignored this and tried the handle, which unexpectedly yielded to my grip. I stepped through the door and spent the next thirty seconds trying to come up with a description of the room I now found myself in that didn't include the word 'jungle', because that would obviously be preposterous.

But there was nothing for it, because I was quite definitely in a jungle: an artificially created jungle on the top floor of a casino hotel in Las Vegas. It was a perfect simulation of a rainforest. There were trees growing up towards the sky – which surely couldn't be right, because the ceiling was too low, unless – ah yes, there was indeed a vast plexiglass dome sitting on top of it all. The humidity had been dialled up to eleven and odd little insects kept buzzing around me. And there were more sounds. Someone had clearly spent a lot of time putting together a realistic jungle soundtrack, because there surely couldn't be all those animals lurking in there. Then again, could there be?

Then I heard a rustle in the undergrowth and glanced down to see something slithering around at my feet. Having spent some time around Patrice's collection of reptiles, even to the extent of playing host to Bertrand the ball python – until he escaped, that is – I considered myself to be reasonably cool about the idea of snakes. However, this feeling only extended to the ones who were firmly enclosed in glass and my reaction on encountering one in the wild was to yelp out loud and begin to develop a sequence of interpretative dance steps that I tentatively entitled 'Shit, there's a snake somewhere at my feet – yes really – a fucking snake!' Fortunately, this had the desired effect of persuading whatever it was to slither off into the undergrowth, but I was left shaking and with a deep desire to get out of there as soon as possible. The presence of something slithery was scary enough in itself, but if there were real snakes in this rooftop jungle, did that mean that the other sounds I could hear originated from real animals as well? No, they couldn't possibly...

I gritted my teeth and kept on walking into the jungle, striding firmly in order to scare away anything else that might be tempted to lurk under my feet. After a few minutes I reached some kind of building, with a sign on the door that read OPERATIONS ROOM. I tried to turn the handle, expecting it to be locked, but it gave no resistance. This briefly presented me with a bit of a conundrum: given that the electrical switch room was kept locked, why was *this* one left unlocked?

A deep-throated growl nearby suggested to me an answer. It wasn't an answer I particularly liked, however. I turned round to see a vast animal covered in orange fur that was pulling itself to its feet and stretching a few metres away from me. This was one of those situations where any internal discussion about the ideal course of action was pretty much superfluous. The only thing to do was move away from it, as quickly and as quietly as possible. Further investigation of this area could wait for another time.

I began to back away in the direction of the door I'd come in through, all the time keeping an eye on the big cat in front of me. Then I turned tail and ran as fast as I possibly could. A quick glance over my shoulder told me that the animal, whatever it was, had decided to join in the fun and, despite the head start that it generously granted me, was now gaining on me with every leap and bound. I turned back to face where I was going and I saw that I was now quite close to where I thought the entrance to the jungle was located. The only problem was that whoever had designed the jungle area had been very particular about making it as realistic as possible, because when I looked for the door through which I had entered, the only thing I could see was a row of tall, ancient trees and creepers blocking my path. Wherever the door might be, I could not see the handle.

Chapter 7

I turned round again, to see that the animal had come to a halt about five metres behind me and was licking its lips as though preparing to get ready for supper. Something went through my head about its vision being based on motion – or was that reptiles? Or dinosaurs? Either way, I tried to keep as still as I possibly could, staring straight ahead, while all the time fumbling with my hands behind me trying to find something that would open the door.

The big cat sniffed the air and it struck me that the idea of keeping still was probably pointless as it was perfectly capable of smelling me whether I stayed stock still in the same position or gave it my best performance of the *macarena*. I could see its train of thought chugging out of the station and beginning to pick up speed as it turned its attention fully in my direction. Come on, come on, there had to be a lever somewh—

Ah, there it was. As the animal reached the point where it was just about to make its final pounce, the door flew open behind me and I tumbled backwards through it, just managing to give it a good slam shut as I fell to the floor. A second later, there was a thump and a very cat-like mewl as half a ton of furry predator smacked the door at the exact point where I'd been standing just a moment before.

There was someone standing over me. It was Don.

'Can I help you, Mr Diego?' he said, although the tone of voice did not suggest any genuine interested in offering me assistance.

'No, I'm fine,' I said as I staggered to my feet and brushed myself down. 'What the hell was that thing? Lion? Tiger?'

'That, my friend, was a liger,' said Don. 'Poppa's a lion, momma's a tigress. Suffers from a certain amount of identity confusion. Tends to be a little bit on the moody side.'

'I got that impression. Is it safe?'

'Depends on your point of view, I guess.'

'But if it had caught me, it probably would have just given me a friendly nip, right?'

'Well, most likely it would have started by tearing off a leg. Slow you down a bit. Then it might have ripped your throat out before moving on to the other limbs. They're pretty methodical critters.'

'Right,' I said, gulping air. 'Right.'

'Still, would have been quick, and that's always a mercy.' Don looked me up and down. 'So, uh, Mr Diego?' he said, 'May I perhaps enquire as to what exactly you were doing in Mr Merritt's liger enclosure?'

Think, Tom. Think.

'I was just looking for a twenty-three mil clutch spanner,' I said.

Jeez. Was that the best you could come up with?

'Uh-huh?' said Don.

'Uh-huh.'

Don thought about this for a moment. 'Twenty-three mil...'

'Clutch spanner, yes,' I said.

'Y'know,' said Don, with exaggerated emphasis, 'I have a feeling you're not gonna find one around these parts. Might need to go to Home Depot or some place.'

'Ah, thanks,' I said. 'That's very helpful. I'll go there, then.' I began to walk towards the corridor that led back to the lift.

However, my way was very soon blocked by an arm. The arm belonged to Don and clearly had no intention of yielding to my tentative pressure.

'You'll be needing transportation,' said Don.

'It's OK, I'll get a taxi,' I replied, trying to duck down underneath the obstruction. The obstruction followed my downward movement.

'I'll take you,' said Don.

'It's fine, honestly,' I said.

'I'll. Take. You.'

I held up both hands in an attempt to defuse the situation. 'OK,' I said. 'OK, cool.'

Don grabbed me by the arm and frogmarched me to the lift. He pressed the button to go down to the basement car park and we descended as quickly as I had come up not long before. When the lift reached the bottom, Don took my arm again and marched me over to a vast black SUV that would have comfortably provided permanent accommodation for a lively family of four plus pets.

'Get in,' said Don.

I got in.

He locked all the doors and I got a strong impression that this wasn't for my own safety.

We'd been driving for around half an hour when it struck me that we had left the suburbs and were now some way out into the Mojave Desert without having seen any sign of a branch of Home Depot.

'Is it much further?' I said.

Don grunted by way of response and continued driving.

'I mean, there must have been one back there, surely,' I said, gesturing towards the way we'd come. 'Or maybe they don't have very good stock at that one, right? Guess this is where a bit of local knowledge comes in really handy.'

'Shut the fuck up, won't you?' said Don.

'OK, OK,' I said. 'It's just, we're wasting valuable time here. That second-tier spofflewack isn't going to fix itself, you know.'

Don sighed and, without taking his eyes off the road ahead, reached into his inside jacket pocket, took out a gun and pointed it at my face.

'Like I said, Mr Diego. Shut the fuck up.'

'Right,' I said. 'Right.' I decided to do as he said.

We drove on in silence for another quarter of an hour before Don brought the juggernaut to a standstill in the middle of nowhere. The place was completely desolate, with only the rocky sandstone outcrops that sprang up on either side of the road for company. If there had ever been a branch of Home Depot here, the sands of the desert had long since washed over and buried its remains.

'OK,' said Don, still aiming the gun right at me. 'Who are you really? 'Cos you know jack shit about electrics, that's for sure.'

I wasn't sure how to answer this, so I decided to go right back to basics. 'I'm looking for my alpacas,' I said. 'Well, they're not exactly *my* alpacas, I just borrowed them, but someone's taken them all the same and—'

'Alpacas?' said Don, genuinely confused by this.

'Yeah, alpacas. You know, like sheep with a ridiculously long neck and—'

'No, that's llamas, man. Alpacas are—'

'They're kind of similar.'

'Don't believe you,' said Don. 'Hold on, I'm gonna google this.' Still keeping the gun pointed in my direction, he grabbed his phone from the dashboard and launched the browser app.

'Well, well,' he said after a couple of minutes. 'Will you look at that. Alpacas *are* just like llamas. Right.' He let out a soft whistle and placed his phone back on the dashboard.

'Yes, they are. So anyway—'

'Who the fuck steals an alpaca?' said Don.

'Two alpacas,' I said.

'Two alpacas,' said Don. 'And you came here to find them?'

'Basically, yes.'

'Well, pardon me, Diego, but that's about as believable as you being a qualified electrician.'

'What?'

'It's bullshit is what I'm saying. Why are you really here?'

'I told you, I—'

Don waved the gun in a threatening manner. 'Y'know,' he said. 'I think I might just shoot you right here.'

'You really don't want to do that,' I said, trying not to sound too nervous. 'Think about the mess it would make.'

Don shrugged. 'Happened before,' he said.

A chill swirled over me as I digested this remark. This human was, despite appearances, turning out to be way more dangerous than Robert J Merritt III's liger.

'Get out the car,' said Don, gesturing with the gun.

I made a quick assessment of my escape options. It was a quick assessment, primarily because there weren't any. If I could make it as far as the rocks, I might manage to hold out for a while, assuming the local wildlife didn't get me first. But there was a long stretch of flat ground to cover before I got there and I would be an easy target if I ran for it.

Don pressed a button on the dashboard and the locks on the doors clicked open. I grabbed the handle and opened the passenger door. I stepped out onto the desert sand at the side of the road and the heat from the desert smacked me full in the face. I looked around. There was still no sign of any nearby cover. Don emerged from the front of the vehicle and pointed the gun at me.

'Start walking,' he said, gesturing towards the side of the road.

'That way?' I said.

'Just do it.'

'Why didn't you just chuck me back in the jungle?' I said. 'Your friend the liger would have made short work of me.'

'This is neater,' he said. 'Buzzards'll strip you clean in a matter of days.'

'That's nice.'

'Just keep walking.'

I kept walking, wondering how in god's name I was going to get out of this one.

'OK, stop right there,' said Don eventually.

'Do I get any last requests?' I said.

'No.'

Then I started to become aware of a kind of buzzing in the distance that seemed to be getting louder with every passing second. I turned towards the road we'd travelled on, and I noticed that a lone motorcyclist was heading in our direction.

'Ah, shit,' said Don, putting his gun away for the moment.

We both turned in the direction of the rider. I hesitated for a couple of seconds, wondering whether if I tried to make a break for it, Don would throw caution to the winds and shoot me anyway, or if he'd be forced to run after me. Don was a big man and I felt there was a decent chance of me outrunning him, although what happened when the bike had gone past was a different matter altogether. Perhaps if I could get high enough on the escarpment to my right, I might be in a good position to throw rocks at him or something.

It was a no-brainer. I had to take my chance now. I took off in the direction of the escarpment, executing a neat zigzag on the way just in case Don was feeling trigger-happy. I reckoned I had about another nine or ten seconds until the biker went past and a world-class athlete could cover a hundred metres in that time. I was not a world-class athlete, but then neither was Don. I heard a shout behind me.

'Stop right there, you little shit!'

I glanced back to see him evidently wrestling with the various options currently presented to him. Then, several seconds too late, he began to run after me. Now that I was pretty sure he wasn't going to shoot, I abandoned my weaving back and forth and instead took off straight for the rocks ahead of me. I was now a decent distance away from Don but I needed to gain some height before my only witness disappeared.

However, I was only just at the point of scanning for the best footholds when the motorbike on the road drew level with me. Damn. I was going to be horribly exposed as soon as they passed me. At that point, things took a turn for the weird as, when I looked again, I realised that the biker had left the main road and was now heading straight across the desert towards Don and myself.

I turned to look behind me and saw that Don had realised what was happening too. I watched in horror as he turned and carefully aimed his gun at the oncoming biker. But he was way too late as the accelerating bike was on him before he could do anything to stop it, and my horror turned to amazement as the person who was on the bike reached down, holding a long piece of metal pipe, and expertly knocked Don's legs from under him.

As Don faceplanted onto the desert floor, the motorbike came screeching to a halt just short of where I was standing, kicking up a localised tornado of sand as it did so. The biker held out a leather-gloved hand.

'Come with me if you want to live,' said a muffled English-accented female voice.

'What?' I said.

'Come with... oh, for fuck's sake, just jump on. OK?'

I still hesitated because my brain was still trying to compute what had just happened. Also, there was something about her voice...

'Ada?' I said.

'Just get on,' she said. 'Quickly.'

For the second time in recent memory, I scrambled onto the back of a motorbike ridden by this mysterious woman called Ada – the only difference this time being that the vehicle in question was a magnificent vintage Harley-Davidson.

We headed back to the road, pausing only briefly to make sure that Don was still out for the count. When we reached the road, we pulled up next to Don's monster truck and Ada climbed off, leaving the engine running. Then she yanked the driver door open and pulled the lever to release the petrol tank cap. Then she walked to the back of the vehicle, took a rag out of her pocket and stuffed it into the tank, before withdrawing it slightly.

'Are you sure you should be doing that?' I said. I didn't fancy being miraculously saved from being shot and left for the buzzards out in the middle of the desert only to get blown to pieces because I happened to be sitting next to the friendly neighbourhood pyromaniac's latest project.

'It's fine,' said Ada. 'I know exactly what I'm doing.' Then she got back on the bike, lit a match and then tossed it in the direction of the petrol tank before accelerating away from the vehicle at full throttle. A couple of hundred metres down the road, Ada stopped the bike and we both turned to watch to see what happened next.

Two seconds later, there was a cataclysmic explosion as the great black beast levitated a couple of feet in the air before discharging its constituent parts in a full three-sixty degree arc. Then there was silence, punctuated only by the rhythmic clattering of the inevitable single hubcap as it spun down to the ground.

'Hate those things,' said Ada.

'I have a number of questions,' I said.

'Later,' said Ada. 'Let's just get out of this place before anyone comes past and starts poking their nose into stuff.'

'What about Don?'

'Who? Oh, the guy with the gun? Well, I guess he has insurance.'

'No, I mean, do we just leave him there?'

'Seems sensible to me. No room for any more on this thing.'

I agreed that this was fair enough.

The journey back to Las Vegas passed without incident and in hardly any time at all we found ourselves cruising through the suburbs and soon we were close to the main strip itself. We pulled up outside the Heartbreak Motel and I dismounted.

'I'd invite you in for a drink,' I said, 'but I'd have to check if it was OK with the cockroaches first.'

'No worries,' said Ada. 'Come to my hotel later and we'll find somewhere a bit more salubrious. Eight o'clock.' She handed me a card. It was from El Gran Pelícano.

'You're staying *there*?' I said.

'Yeah. Is that a problem?'

'Bit of one. The guy who just tried to kill me came from there. I was doing some work for him up in the Presidential Suite.'

'Wear a disguise or something. They probably won't remember you.'

'That's easy for you to say but—'

'Yeah, it is. Look, I gotta go.'

'OK, but how will I find you?'

'Just ask at reception.'

'But who do I ask for?'

'Ada,' she said. 'Ada Vavasor.'

Chapter 8

My head spun as I watched her accelerate away up the street. Up to this moment I had been under the impression that there weren't any Vavasors left alive. The twin mathematicians Archimedes and Pythagoras were long gone and their younger brother Isaac had been killed less than a year ago by one of the Fractal Monks and for all we knew, his body was still in their monastery back in Meteora. The idea that there could be another generation of Vavasors was a possibility that I had completely failed to anticipate.

To say that I had mixed feelings about all this would be to understate things by some considerable distance. My feelings on the subject had basically been beaten firmly with a fork before being passed through a professional quality kitchen blender and then finally tossed into the hopper of an industrial centrifuge. If it hadn't been for the Vavasors, I would almost certainly have never broken up with my ex-girlfriend Lucy – but, then again, I would almost certainly have never got back together with my childhood sweetheart Dorothy. Then again, if it hadn't been for the Vavasors – or to be more accurate, the Vavasor papers – or to be even more accurate, my accidental incineration of the Vavasor papers – I probably wouldn't have broken up with Dorothy again either.

But somewhere in the middle of all this, a tiny voice was trying to make itself heard: the tiniest rodent squeak of a voice that was trying to convince me that somewhere there might be another copy of the Vavasor papers and that Ada might just be the person who'd know where to find it. And if I could recover the Vavasor papers, perhaps I could recover my relationship with Dorothy. Forget trying to get the alpacas back, *this* was the reason fate had brought me to Vegas.

Then again, what in god's name was Ada doing here? And how had she found me? What was Ada's agenda here, after all? Come to think of it, what had Ada's agenda been at the Luxxy Duxxy warehouse?

Something told me that getting my hands on the goods wasn't going to be a case of simply going up to her and saying 'Hey, Ada, can I come up and see your papers some time?' I needed to prepare myself for our next meeting. I needed to take a shower, relax for an hour or so and gather my thoughts. I smacked my hands together and, with a muttered 'Come on!' to myself, I walked back into the lobby of the Heartbreak Motel.

'Evening!' I shouted to the receptionist. She responded by regarding me with a deeply suspicious eye, as if I'd been caught attempting to have unnatural relations with one of my cockroaches.

I walked back to my room and, as soon as I opened the door, I knew that something was wrong. I wasn't a particularly tidy person at the best of times, but this was a thoroughly systematic mess. Someone had been in here and rummaged through my stuff. I'd been very careful not to give my address to Jolene, although to be honest, she hadn't actually asked for it. The job was strictly cash in hand, no questions asked. Had I been followed then? But who would go to the trouble of following me? Don was presumably still out in the desert, coming to with a very sore head and wondering what had happened to both his car and the phone he'd left on its dashboard. There was absolutely no way he would have had time to get back to the city and send anyone after me.

What had they been after anyway? I'd brought nothing apart from my washbag, a couple of books and a change of clothes and nothing seemed to be missing. Also, what would have happened if I'd been there? What if they came back? Was I in danger? So much for a relaxing shower and a nap. I had to get out of there.

I quickly got changed. As I left the room and closed the door behind me, it struck me that there didn't seem to be any evidence of damage to the lock. I went back to the lobby, where the receptionist had abandoned any speculations about my proclivities in favour of a game that she was playing on her phone.

'Excuse me?' I said.

'Uh-huh,' said the receptionist, not looking up.

'Someone seems to have been in my room,' I said.

'Uh-huh?'

'Yes, I'm afraid so,' I continued, trying very, very hard to remain calm. 'Do you perchance have any idea why that might have come to pass?'

The receptionist shrugged and went back to her phone.

'I take it you have spare keys?' I said.

'Uh-huh.'

'So did you perhaps lend them to anyone?' I emphasised the word 'lend' with air quotes and immediately felt embarrassed that I'd stooped so low.

The receptionist looked up at me, clearly irritated that I was interrupting quality gaming time. Then she gave a vague wave of her hand towards the area under her desk. I leaned over and saw a neatly labelled array of keys to every single room in the motel.

'Ah,' I said. 'So if you happened to be away from your desk for any reason, someone could just come along, check your reservation book and pick up the key to anyone's room, right?'

'Possibly,' she said. I was so shocked at her giving me something close to a straight answer that I almost forgot the follow-up question.

'OK, then,' I said, pointing up at the security camera on the wall behind her, 'presumably the camera up there will have a record of who it was?'

The woman shook her head. 'Not working,' she said.

'What?' I slapped my hand on the counter. The receptionist tapped the 'Don't Be Cruel' sign on the desk and then pointed to a notice on the wall behind her next to *Blue Hawaii* that said that the Heartbreak Motel had a policy of zero tolerance of aggression towards their staff and that any guest engaged in such behaviour would be liable to immediate eviction with no refund.

'OK, OK,' I said, holding my hands up, 'I'm sorry. Just a bit frustrated, that's all.'

The receptionist didn't bother making any further comment. Her phone suddenly gave out a triumphant sequence of beeps and she made a half-hearted attempt at punching the air before returning to her game. Trying very hard to keep my homicidal thoughts to myself, I walked out into the Vegas dusk. I wasn't going to get any further with her.

I still had a couple of hours to kill and I wanted to kill them as quietly and as far away from El Gran Pelícano as possible, so I found a small bar in a side street away from the main strip. I took my seat at the bar and absorbed the conversations going on around me. Behind me and to my right, a couple were earnestly engaging in a desperate first-date search for something in common to talk about and had finally ended up comparing the different shades of amber in the various glasses of beer in the establishment. I guessed there were relationships with a less auspicious origin story, but on reflection probably not that many.

Behind me and to my left, an affair of greater longevity was drawing to its inevitable close, judging from the detailed list of complaints that the woman was reeling off and the apparent indifference of her partner to the concept of addressing any of them. Half an hour in, the argument had built towards its inevitable finale which involved her calling him a lazy good-for-nothing money-wasting poker-obsessed loser, emptying the contents of a full glass over his head and storming out. The target of her abuse sat there for a few seconds with beer dripping down his face before letting out a deep sigh and following after her. This was clearly not the first time this had happened.

There was a long, thoughtful silence in the bar. Then the shaven-headed guy on the bar stool to my right turned to me and said, 'I don't like you.'

'I'm sorry?' I said.

'I said, buddy, that I don't like you.' He prodded my chest for emphasis.

'I—' I began. I wasn't that keen on him either, come to think of it, but I wasn't intending to make a thing of it.

'Leave him alone, Carl,' said the barman. 'Maybe time to go home now?'

'Bastards,' muttered Carl, easing himself to a standing position. 'They're fucking bastards, every one of them!' He chucked a few dollars at the barman and picked out a path towards the door.

'Another beer,' I said. The barman nodded and refilled my glass.

'You in town for the convention?' he said.

'Do I look like it?'

'Well, you sure gotta funny accent.'

'Where I come from, we all talk like this.'

'Well, it's kinda cute, I guess. Bet the ladies love it.'

'Oh yes,' I said. I didn't often get the chance to flaunt my prowess with women. 'Listen,' I said, wondering if I could seize the opportunity to capitalise on this moment of ersatz masculine intimacy, 'what's the deal with Robert J Merritt III?'

'What, the Gran Pelícano guy?'

'Yeah, that's the one.'

The barman leaned in closer. 'Well,' he said, 'apparently he has his own artificial jungle on the top floor of his hotel. Can you believe that?'

Weirdly, I could. I said as much, although I left out the minor detail that I'd actually spent some quality time in there with one of the occupants.

'And the word on the street is that everything in his casino is completely rigged. But of course no one's managed to prove it. Or at least—' here his voice dropped to a whisper '—no one's managed to prove it and stay alive for longer than another hour or two.'

'Really?' I said.

'Yeah, really. The things I could tell you.'

'Tell me, then,' I said. 'Go on.'

The barman stood up and gave a dismissive wave. 'Nah,' he said with a laugh. 'I'd have to kill you afterwards myself.'

'Well, that wouldn't be a good idea,' I said, 'now would it?'

The barman cackled and moved on to serve another customer. I sat and sipped my beer for a few minutes, wondering what someone like Merritt would have to do with a disused plastic duck factory on the outskirts of Basingstoke. Didn't he have enough to do with his casino? If the operation was as bent as my new friend claimed it was, presumably he must be raking it in.

I finished my beer and checked my watch. There was still plenty of time to go until eight o'clock, so I signalled for the barman to pour me another one. He was about halfway through that when he gave me a conspiratorial wink and said, 'Don't look now, but there's a couple of Merritt's goons just walked in.'

A shiver ran down my spine. I snuck a glance behind me and to my horror I saw Jolene entering the bar, followed by a heavily limping Don. Shit. I was trapped. It was surely only a matter of moments before I was identified, dragged out and disposed of properly this time.

'Excuse me one moment,' I said to the barman. 'Could you perhaps tell me the way to the bathroom?'

'Sure, buddy,' he said, putting the beer down in front of me. 'Over there to your left.'

I took a quick swig from my beer, left a handful of notes on the counter and scuttled off to my left, ducking my head low to avoid being recognised. As soon as I reached the bathroom, I slammed the door behind me and quickly checked out the cubicles. There was a frosted glass window above the middle one that looked as if it might open sufficiently to let me through. I locked the cubicle door, climbed up onto the seat and gave it a hearty thwack with my forearm. The window failed to budge, although one of the panes of glass in it obliged by tumbling out and crashing noisily onto the alleyway outside.

Another solid thump and a couple of the other panes fell clear. If I could just shift the frame itself, I could escape. Then I realised that in my haste I hadn't actually opened the lever. I gave it a good tug and this time the window flew open. I clambered up and launched myself through it, promising to myself that I'd come back some time in the near future to pay for the damage.

I was halfway through the window when I realised that one of my belt loops was caught on some part of the window mechanism and had no intention of letting me go without a fight. I wriggled and squirmed for several useless seconds but there was no way I was going to get out of this without taking some kind of drastic action and it was only a matter of time before someone happened to wander down the alleyway and wondered what was going on.

I twisted around and grabbed hold of both sides of my trousers. Then I took a deep breath and tugged as hard as I could. There was a brief moment when it felt as if the very fabric of time and space itself was deciding whether to give way or not, then the button pinged off, the fly unzipped all the way down and I tumbled out of my trousers down into the alleyway head first, kicking off my shoes as I went.

As I fell out onto the pavement, I somehow managed to miss most of the broken glass from the window pane that I had helpfully smashed a couple of minutes earlier, although when I stood up, I realised that my stockinged feet were now balanced nicely on some of the shards. I shifted my position slightly, and picked the larger and more vicious pieces out as best I could, hoping that my socks would cushion my feet from the smaller ones that remained. Then I dusted myself down and unhitched my trousers from where they had got caught. I reached back in through the window to retrieve my shoes from where they had landed on top of the cistern and at that point I heard the first shout of 'Stop right there!' I saw Jolene at the end of the alleyway behind me and realised that I wasn't quite out of the woods yet.

Carrying my scrunched-up trousers in one hand and my shoes in the other, I set off in my socks towards the opposite end. Three-quarters of the way there I glanced back to see Jolene in hot pursuit, but where was Don?

I soon found out. As I approached the point where the alleyway spilled out onto the strip, Don emerged, hobbling unsteadily towards me. However, my momentum was now such that there was no alternative for me but to hurtle on towards him and a few seconds later I collided with Don, sending him toppling over like the front pin in a game of pub skittles. I bounced off him and continued on my way towards the strip.

As I reached the end of the alleyway, I took a right turn and headed on towards El Gran Pelícano, wondering how much further I dared go before risking a stop to make myself decent again. I wasn't sure what kind of view the Nevada legal system took of semi-naked men running barefoot down a public thoroughfare and I was in no hurry to find out one way or the other. I slowed down for a moment and looked back at where I had come, but there was no sign of either Don or Jolene.

I stopped, put my shoes down on the pavement and, as unostentatiously as possible, pulled on my trousers. I zipped up the fly and almost as soon as I had done this, they fell down again. I zipped up once more and tried breathing out. But it was no use: as soon as I inhaled, down they went. I was clearly going to have to get wardrobe assistance from someone as soon as possible, because I couldn't risk going back to the motel again. I pushed my feet into my shoes and walked on holding my trousers up, towards my meeting with Ada.

Chapter 9

I recognised Ada immediately as she was still wearing her black leather catsuit. She was, however, no longer wearing any form of head covering, be it balaclava or bike helmet. Her hair was a startling shade of bright red and her lipstick was bright red to match. She was wearing a pair of entirely unnecessary dark glasses. It was nice to see her face at last, even if the eyes were still hidden from view.

'You didn't need to change, you know,' I said.

'I'm not sure you're in a strong position to criticise my attire,' she said, peering at me over the rim of her glasses and giving me a sardonic smile.

'Long story,' I said. 'You don't happen to have a safety pin, do you?'

She rummaged in her bag and produced one with a theatrical flourish. I thanked her and cinched my trousers firmly around my waist as discreetly as I could. It was good to have both hands free again.

'Right,' said Ada. 'Shall we go? I'd suggest we eat here but apparently the enchiladas are just asking for trouble.'

'I would be very happy to,' I said. I really didn't want to spend any more time inside El Gran Pelícano than was necessary. It was true that, apart from Jolene and Don and – I guess

– Larry the Liger, no one really knew me there, but it still made me feel nervous.

'I think I know just the place to go,' said Ada, leading the way.

This turned out to be a small Italian restaurant in a side street a few blocks down from El Gran Pelícano. As we walked there, I kept glancing over my shoulder for any signs of Don or Jolene, but it seemed that they had finally given up on me for the day. I took my seat and began to feel the closest to safe that I'd felt all day. I finally felt able to relax. We ordered drinks and then spent some time examining the menu.

'So, Tom,' said Ada. 'I guess my first question is what on earth are you doing here?'

'I have a similar question,' I said, 'along with a side order of "Are you really one of the Vavasors?" and a sprinkling of "How, in god's name, did you know I was being kidnapped and taken out to the desert to be killed?"'

'Well, I'm happy to start by answering the last one, I guess,' she said. 'I was sitting in the lobby of the hotel and I recognised you immediately when you came in and got shown the way to the big man's private lift. I remember thinking to myself, "That's probably not going to end well". So I nipped down to the basement and waited for you to appear, and right on cue you did, arm-in-arm with your very good friend Don.'

'He made it back, by the way.'

'Well, that's nice.'

'Got a bit of a limp, though.'

'It's the least of what he deserved.' Ada paused and took a sip of her spritzer. 'OK, my turn. What on earth are you doing here?'

'Same as I was when I first met you.'

'Which is?'

'Looking for my alpacas. Well, not *my* alpacas, in fact. Someone else's. That I borrowed a while back.'

'And what happened to these,' she paused, 'alpacas?'

'Someone stole them,' I said.

'And you thought you'd find them in Las Vegas?'

'Well, not exactly. More a case of following the only lead I had.'

Ada stared at me for a moment, her head on one side. 'That's not the whole story, though, is it?'

'Isn't it?'

'This isn't a game, Tom.'

'You sure?'

At that point, the waiter came to take our order. When she'd gone, I remembered something important.

'By the way,' I said, holding up my wine, 'thank you for saving me from Don. All things being equal, I quite like being alive.'

'It's generally a good thing,' said Ada, chinking my glass. 'To be honest, I only did it because I think you might be useful.'

'What for?'

'I'm not even sure yet.'

'That's not terribly helpful.'

'I know. Look, something's about to happen. Something involving Robert Merritt.'

'I imagine things happen around him all the time,' I said. 'The sort of person who has his own private jungle in the penthouse suite of his hotel is exactly the sort of person that things happen around.'

'Fair point,' said Ada, leaning forward. 'How do you know about the jungle?' she added.

'I've been in it. Got chased out by a liger.'

She shook her head slowly. 'From what I saw of you in action at Luxxy Duxxy, that seems pretty consistent with your style.'

'That's a bit mean.'

'Is it?'

'OK,' I said, anxious to seize control of the conversation again. 'Tell me about the Vavasors. Actually, don't tell me. I already know far too much about them.'

'Ooh, are you a Vavvy?' 'The Vavvies' was a colloquial way of referring to the online community of Vavasorologists – a weird mix of obsessives, fanatics and conspiracy theorists who coalesced around the mystery of what had happened to the Vavasor twins, Archimedes and Pythagoras.

'I might have been once,' I said.

'In which case something tells me you aren't quite what you seem, Tom. Because any self-respecting Vavvy would have already asked me if I knew what had happened to the Vavasor papers and you haven't muttered a word about them.'

'I was just about to,' I said.

'No you weren't,' said Ada. 'So that means you're either not a proper Vavvy at all, or – and this is the interesting possibility – someone who knows a lot more than they're letting on.'

At this point, I was very tempted to show off by admitting to being in the latter camp, but I decided to keep my powder dry for the moment. I said nothing.

'Then again,' continued Ada. 'Looking at you, I find it very hard to believe that *you* would know anything about my family that isn't already firmly in the public domain.'

'That's a bit unfair.'

'No, it's a compliment. Honestly. You don't look like the run-of-the-mill Vavvy at all. Way too normal.'

The food arrived at that point, interrupting the flow of our conversation. We ate in silence.

'This isn't at all bad,' said Ada eventually.

'It's certainly better than anything I ate yesterday.' I took another mouthful, considering what to say next. 'So where do you fit in to the family tree?'

'My father was Isaac Vavasor,' she said.

'Oh,' I said. We were sailing very close to some very turbulent waters now.

'Yes. You may have heard what happened to him.'

'I'm not sure of the details,' I lied. I made a snap judgement that this would have been an inauspicious time to introduce the fact that I'd actually *been* there at the moment of Isaac's murder at the hands of one of the Fractal Monks. There would probably be some awkwardness later, but I needed to know more about Ada's agenda before I undid the tie on the old kimono. 'To be honest, everything I know is second-hand rumour anyway,' I added, immediately resolving to come clean at some point. In the not too distant future.

'I mean, we weren't close, not at all. In fact, we hadn't spoken to each other for several years, but it was a bit of a shock nonetheless.'

'Go on,' I said, trying to sound as sympathetically ignorant as possible.

'Apparently he was in Macedonia, visiting some monastery or other.'

I came so close to blurting out the word 'Greece' that the paroxysm that I conjured up to cover my tracks almost caused me to choke on my pasta.

'Are you OK?' said Ada.

'Imfine,' I gasped. 'Imfineimfineimfine.'

'Try drinking some water out of the opposite side of the glass, Usually works a treat. If you'd prefer me to try the Heimlich Manoeuvre, just raise your right hand.'

'Illgiveitago,' I said, keeping my right hand duly lowered. My attempt to drink from my glass in the manner that she had suggested resulted in me spraying water over most of the table in front of me. However, it did have the desired effect of restoring me to normal.

'Anyway,' said Ada. 'Where was I?'

'Macedonia,' I said.

'Yes. Well, apparently, he was visiting this monastery and this deranged monk stabbed him for no good reason.'

'Bloody hell!' I said. 'That's awful.'

'Well, yes. Although, as I say, we weren't on the best of terms, what with one thing and another.'

'Can I ask why?'

'You can ask,' said Ada. 'But I won't necessarily feel obliged to give you an answer.'

'Fair enough. Why haven't I heard of you, anyway?' I said.

'I've been out of the loop for a while.'

'Can I ask you about that?'

'No.'

'Can I ask you what you're doing in Vegas?'

Ada finished her pasta and leaned back for a moment, considering my question. 'OK, let's say "yes" for now and see where it takes us.'

'Right,' I said. 'What are you doing in Vegas?'

'Well, it all started when I bought a bike.'

'A bike?'

'Yes, a motorbike.'

'Ah.'

'An electric one, to be precise. Very nice it was too.'

'You seem to like your bikes,' I said.

'Given that they've got you out of scrapes on both sides of the Atlantic, you should like my bikes too, Tom.'

'Oh, don't worry, I do. I am exceptionally fond of your bikes.'

'I mean, the Harley is hired, obviously, so it's not really my bike. But anyway, I digress.'

'You were telling me about your electric bike.'

'Yes, I was. Lovely piece of engineering. Massive torque but the smoothest ride of your life. Gorgeous. Trouble was, I'd had it about a week when all of sudden one day it refuses to start. Battery fully charged, but absolutely nothing doing.'

'So what did you do?'

'Well, I was about to take it back to the place I'd bought it from when I received an email. From my bike.'

'Uh-huh.'

'Yep. An email telling me that, unless I paid ten grand in Bitcoin, my bike would never move again. It was effectively bricked.'

'You're kidding.'

'Nope. So I phoned up the dealer, told them the story and they came and took it back to their workshop. Three hours later I get a call telling me I'm basically stuffed, because somehow not only has my nice new bike been bricked, but in the process my warranty has been violated and no one can do anything about it. Also, at around the same time, I received another email from the bike telling me to stop twatting around trying to break into it and just pay up.'

'Shit. So what happened next?'

'Well, I brought the bike back home, stuck it in the garage and went back to using the evil fossils until I could find a way to get the bloody thing working again.'

'You have a garage full of bikes?'

'One or two. Three. OK, half a dozen. It's my one indulgence. Well, one of my indulgences. Look, when you grow up in a family like mine—'

'It's OK,' I said, holding up my hands. 'Don't worry about it.'

'Anyway, a couple of months after all this happened, I was chatting to a programmer friend of mine and he was wondering if it might be possible to trace where the bug had come from. But surely we were too late for that, I said. Not necessarily, he reckoned. Those threatening emails that the bike sent if anyone tried to break into it and remove the bricks, were all very nice and theatrical, but chances were, they weren't actually coming from the bike. Whatever virus they'd managed to inveigle onto it wasn't going to be that clever. The emails had to be coordinated from somewhere central.'

'Aha!' I said. 'Nice work.'

'Yeah, he's a clever bod. Not exactly house-trained, but he has his uses. Anyway, my chum comes round, we make sure the bike is online and hooked into my network and he sets this app running on my laptop.'

'Wireshark?' I said. Something had just pinged in my brain about the time Ali had tracked down the location of the Tulpencoin crypto server on Channellia.

'Yes,' said Ada with a quizzical frown. 'That's the one. Are you sure you don't know more than you're letting on?' I said nothing. She continued to look at me for a few seconds and then continued. 'So we start trying to hack into my bike, same as the garage would have done. We hook up to the various USB ports and try to prod it into action until eventually it spews out one of its angry emails. My mate stops his Wireshark thing, takes a look and announces that he has an address.'

'Brilliant.'

'Yeah. Another drink?'

'Sure. They're on me, by the way.'

'So they should be.'

We called the waiter over for a top-up and Ada continued. 'So my techie chum did one of those – what do you call it? I think he called it a "whois" – things on the address and finds out that it's owned by a defunct manufacturer of bath ducks with a factory on the outskirts of Basingstoke.'

'They didn't even bother to change internet service provider?'

'No point. Hosting probably still paid up for the duration of the contract, so they're just squatting there for the time being. Then when it's finished they'll move on somewhere else. Plenty of companies going bust these days.'

'Bloody hell.'

'Yeah, bloody hell. So I guess that explains why I'm in Vegas. Basically I want to get my bike fixed.'

'And I want my alpacas back.'

'I still don't understand that one.'

'Neither do I.'

We both sipped our drinks in silence.

'Can I just ask,' I said eventually, 'why you didn't just go to the police?'

Ada lowered her sunglasses for a moment. 'You really need to ask that question?' she said.

'I guess not.'

'The thing is, we both know that the police aren't likely to be interested in some convoluted issue with a dodgy motorbike that may or may not even turn out to be a crime, and even less interested in the disappearance of a couple of alpacas.'

'Exactly.'

'Or at least that's the story we tell ourselves,' continued Ada. 'The real truth is that we both fancy ourselves as private investigators. Am I right?'

I didn't say anything.

'I'll take that as a yes,' she said.

'So what's the plan?' I said.

'Not a clue. Tell me what you found in the penthouse before you got thrown out and maybe that'll give us a pointer or two.'

I described everything I'd seen up there, which wasn't an awful lot, now I came to think about it. Ada wasn't interested in the floor coverings and the gold trimmings, either, so pretty much all I was left with was the room containing the piece of equipment I was supposed to be fixing.

'So no computer terminals at all?' she said.

'Nope. Just a load of racks full of units with blinking lights on the front. Oh yes, some of them were labelled.'

'Labelled?'

'BTOKE,' I said.

'Btoke?' said Ada.

'Yeah, Btoke. B – T – O – K – E.'

'Oh, for fuck's sake, Tom. Basingstoke! It's something to do with Luxxy Duxxy!'

'Might be I suppose.'

'There's no "might be" about it. Probably some kind of networking hub. Can you remember any of the other ones?'

'There was a SPORE,' I said.

'Ooh,' said Ada. 'Bet that's Singapore. Any more?'

'Nope,' I said. 'Sorry.'

'Doesn't matter. OK, so we know where the main hub for the network is. You're definitely certain there wasn't anything that looked like a computer in there?'

'Ninety-nine point nine per cent, no.'

'Got to be somewhere else up there, though. No point in sticking all the networking stuff in the bad guy's HQ if his processing power's somewhere else.'

I suddenly remembered there was something else I'd forgotten to mention. 'I have a horrible feeling I know exactly where it is. There's this other building you see. In the jungle.'

…oman shook her head, tutted and stalked off in the
…of the ice machine. We both broke down into uncon-
…laughter. Ada then spent several minutes trying to
…o her room by sliding her card down the gap between
…and the jamb.

…pe,' I said.

…pe,' she agreed and then slid it into the slot next to the
…and miraculously the door opened. I followed Ada into
…om and then she closed the door behind us. She sat on
…ed and I sat on a chair. Then with a dramatic flourish, she
…oved her bright red hair to reveal a severe buzzcut. She kept
…shades on, however.

'Woah,' I said. She looked even more striking without hair
…n she did with.

Ada smiled enigmatically, then she got up and went over to
…e minibar before taking out a couple of miniatures of whisky
…nd some packets of peanuts. She tipped the contents of one
…f the bottles into a glass and handed it to me, before filling
another one for herself. Then she tossed a packet of peanuts in
my direction.

'You know what?' she said. 'Seeing as we're not going to
have sex, we should do something really sinful instead.'

'Like what?'

'Spend the night gambling.'

'Have you seen it down there? It's all slot machines. It's all
a bit depressing.'

'OK, so we find the roulette tables. Or blackjack. Or
whatever else they do down there.'

'I've no idea how to play blackjack. We'd get rinsed.'

'What do you know how to play, then?' said Ada. She
opened the bag of peanuts and took one out. Then she threw it
in the air and attempted to catch it in her mouth. She missed by
several inches. 'Shit,' she said, rummaging around on the floor
trying to find where it went.

Chapter 10

'So what you're saying is that the computer system that we need
to get hold of is guarded by Larry the liger?'

'Yeah, basically.' We were walking back towards El Gran
Pelícano. I was realising that I'd had more to drink than I'd
intended and I suspect that Ada was in a similar state, judging
by the number of times we bumped into each other as we
continued on our way.

'So how do we get past him?'

'Or her. Could be Larrietta.'

'Not a proper name.'

'Leonora, then.'

'Leonora would be a full-blown lion. Leo, see. Not a liger.'

'Well, doesn't matter. Whatever it's called, it's got razor-
sharp teeth and probably a massive appetite, and we've got to
get past it if we're going to find out what's happened to your
bike. And maybe my alpacas.'

'Any zoos round here?' said Ada. As she said this, she
stopped suddenly, grabbed my arm and turned towards me. I
stopped walking too.

'Zoos?' I said.

'Yeah, zoos. We need someone on the team who can keep
the liger happy.'

'Team?' I said. 'Are we recruiting for Vavasor's Eleven or something?'

'Maybe we are. Are you having second thoughts?'

'Of course not.' I was, of course, having loads. This was getting more and more stupidly dangerous by the minute and there wasn't any guarantee whatsoever that I was going to get any tangible benefit from it. For all I knew, Dolores and Steven were already well on their way into the food chain, either back in England or in whatever other country they'd ended up in. But I was also sufficiently drunk for these second thoughts not to override the first ones, which were along the lines of 'Hell, yeah, show me the way to the heist!'

We started walking again.

'OK,' Ada was saying, 'so we're going to need a big cat expert. How much do you think we need to pay someone like that?'

'To get involved in a potentially criminal enterprise?'

'It's only criminal if we get caught before we manage to expose Merritt.'

'Actually, it would be easier if it was a proper criminal heist. Could offer them a share of the proceeds.'

'Anything up there we could nick?' said Ada.

'Plenty of gold fittings, but we'd need to unscrew them from the walls.'

'Won't have time for that.'

'Also, how do the three of us get up to the penthouse anyway? They'll be on the lookout for me, for one thing.'

'Disguise? You could shave your beard off.'

'Are you kidding? It's taken me months to grow that.' It was my break-up beard after everything had gone down the tubes with Dorothy. I was quite sentimental about it. 'On the other hand,' I said, 'if we're talking about forming a team, maybe we could recruit someone who works for Merritt?'

'Know anyone?'

'Might do.' I was wondering ... or Jesús to moonlight for us.

'Great, looks like we might ... team, then. Anyway, here we are a... out her hand. 'Lovely to have the cha... you.'

I took her hand and shook i... something. 'Look,' I said. 'I've got a bi... accommodation.'

Ada tipped her sunglasses forward an... the eye. My mild wooziness cleared in an i...

'No, hang on,' I said. 'I wasn't... it's n... look, I'll explain.' I described to Ada how m... ransacked and how it probably wasn't going t... to go back there. She continued to look at me f... seconds after I'd finished talking.

'OK,' she said finally. 'You can kip on my floo... Tom. But in case of any misunderstandings, I'd like... clear that this invitation only extends as far as the flo... further up. We've both had a few drinks and I woul... anything to happen that we might later regret.'

'Of course,' I said quickly. 'The thought had never... ... hadn't. I'm sure it hadn't.' I was absolutely sure it hadn't.

'Besides,' she added, pushing her glasses up again, 'I n... have sex on the first date.'

'Right,' I said, supressing a gulp. 'Right.' I followed her int... the lift up to the eleventh floor. As we exited, I noticed one of the other guests coming out of one of the rooms opposite Ada's. She stopped and looked at us and for some reason I assumed she was making some kind of insinuation about us.

'It's all right,' I called out. 'She says she doesn't have sex on the first date.'

Ada sniggered. 'He's right,' she said. 'Never do. Never never never.'

'Played a bit of poker,' I said. 'When I was at uni. Used to lose every time. Never worked out why until Dorothy explained it to me.'

'Dorothy?'

'My ex. Mathematician.'

'Oh god.'

'Why "Oh god"? I thought you grew up in a mathematical family.'

'That's precisely why "Oh god", Tom,' said Ada. She took another peanut out and tried to catch that in her mouth. She missed that one as well. 'Bollocks,' she said. 'OK, so what's the secret to poker, clever clogs?'

'Well,' I said, racking my brains to try and remember what it was that Dorothy had told me. 'It's all to do with Bayeseseseses' theorem.' I looked at my glass. It was almost empty. Ada went to the minibar, took out two more miniatures and threw one at me. I caught it and poured the contents into my glass.

'Bayeseseseses' theorem?' said Ada.

'Just the one Bayes, I think. With an apostrophe at the end. Bayes'.'

'OK.'

'So basically it's all about conditional probability,' I said, feeling quite impressed that the phrase had popped into my head right on cue. 'The probability that event B happens given that event A happens. So, if you think about the probability of both A and B happening, it's going to be the probability that A happens, multiplied by the conditional probability of B happening given that A happens.' I paused for a moment. 'Shit, did I get that right? I think I did.'

'Do you want to write that down, though, Tom?' said Ada. 'I'm struggling a bit here. Enjoying the sexy talk, though.'

I got up and found a pad of writing paper in the drawer of the dressing table along with a Hotel El Gran Pelícano ballpoint pen. Then I took the pen and scribbled this on the pad:

$$P(A \cap B) = P(B|A) \times P(A)$$

Then I passed it over to Ada.

'Uh-huh,' she said, handing the paper back to me. 'What's the upside-down U thing?'

'It's the sign for "and". So it's the probability that A and B both happen.'

'And the vertical bar?'

'It's the sign for "given". So it's the probability that B happens given that A also happens.'

'OK, makes sense. I think.'

I tried to remember what came next. Ah yes.

'So the thing is, the probability that both A and B happen is the same as the probability that both B and A happen.'

'Obviously.'

'So we can do this,' I said.

$$P(A \cap B) = P(B \cap A)$$

'Uh-huh,' said Ada. 'So what?'

'Now this means we can do this as well.'

$$P(B|A) \times P(A) = P(A|B) \times P(B)$$

Ada stared at this for a long time and then nodded.

'Got a feeling Uncle Archie might have tried to explain this to me at some time,' she said. 'Didn't go in at the time but it's beginning to make sense now. Either that or I'm just enjoying the symbols floating around on the page.'

'OK, we're nearly there,' I said. 'Let's just rearrange this a bit.'

$$P(B|A) = P(A|B) \times P(B) / P(A)$$

Dorothy would have been so proud of me. Ada looked at it and shrugged. 'What's this got to do with poker?' she said.

'Ah,' I said. 'Just give me a moment to work it out.'

'Sure,' said Ada. 'All the time in the world.' She leaned back on the bed and tried the peanut trick again. This time she caught it in her mouth and she bounced up and down on the bed in triumph, waving her arms in the air.

'OK,' I said. 'I think I've got it.'

'Good,' said Ada, sitting up and trying to look very serious.

'No, hang on,' I said. 'Lost it again.'

''S'OK. Try again. In your own time.' Ada tried to repeat her peanut success with two at once and missed both. 'Bugger,' she said.

'Right,' I said. 'Think I've got it now. In poker, event B is that the player I'm up against has a winning combo. Event A might be something sensible like, I dunno, he raises. Or it might be some habit – some kind of "tell" – like he sticks his index finger in his left ear and waggles it in a clockwise direction. Now, by looking at my cards and the ones on the table, I should be able to work out what the baseline probability of event B is. But what if my opponent has just waggled his *left* ear?'

'In a clockwise direction, don't forget.'

'If I know the probability of event B given event A, I've got a much better idea of how good his hand is. Now the thing is, I'm really clever and I've been observing his ear-waggling all night, so I know precisely what the baseline probability of him waggling his ear is. And I've also observed what the probability is of him waggling his ear if he happens to go on to have a winning hand. So I take my baseline probability of him having a winning hand and multiply it by the probability of him ear-waggling if he's got the winning hand and then divide it by the probability of him waggling his ear.' I paused for a moment, going over what I'd said. 'I think that's right, anyway.'

'And what exactly are the chances of you managing to hold all that information in your head, Tom?'

'Roughly zero,' I said. 'But it's the thought that counts.'

Ada launched a peanut into the air towards me. I tracked it as it described a perfect parabola and then landed about a foot short of my mouth.

'You're right,' said Ada. 'We'd get absolutely rinsed. I'm going to bed.'

She stood up and began rummaging in the wardrobe for spare pillows and bedding. She threw a random selection of stuff on the floor in front of me.

'That do?' she said.

'Guess so,' I said.

'Good. Time to get some sleep. Lots to do tomorrow. Plans to draw up. Places to go. People to see.'

I arranged my improvised bed on the floor and lay down on it. It could have been a lot worse. What a day it had been. I was quite glad that I was drunk because if I'd been sober I would have spent far too long going over what had happened and trying to work out some kind of logic to explain it all.

I slept well considering the circumstances, but I awoke with a massive thumping jackhammer of a headache and the feeling that every last drop of the water that made up sixty per cent of my body mass had somehow contrived to find its way to my bladder. I hauled myself upright, eyes half-closed, and staggered to the bathroom to relieve myself, just about managing to stay upright by pressing one hand against the wall above the toilet. I came back into the room to find Ada fully dressed and pottering about making coffee. She'd decided to go for a bright green wig today, with matching lipstick, although the rest of her outfit was identical to the day before. I sat back down in my chair, with my blanket still round me.

'Morning,' she said.

'Ugh.'

'Sleep well?'

'Ugh.'

At this point, she flung the curtains wide and the dazzling desert sun poured into the room.

'Now that,' she said, 'is quite a view, is it not?'

'Ugh.' I refused to move from my chair.

'Ugh?'

'I meant,' I said, 'that I'll trust your judgement on that.'

'See? Sentences *are* possible, Tom.'

'Ugh.'

Ada handed me a cup of coffee.

'Drink.'

'Don't want to.'

'Drink.'

I managed to take a few sips and within a minute or two, I was beginning to feel human again. Or at least somewhere scrabbling around in the lower branches of the order of primates. I managed to drink the rest of the cup and the caffeine finally worked its magic.

'OK, I seem to be alive,' I said. 'What's the plan for today?'

Ada turned back from the window, still holding her coffee cup in her hands. 'I've found a zoo,' she said.

I remembered now. Everything from the day before that I'd been holding back suddenly came rushing in.

'So we're really going to do this?' I said.

'Oh yes,' said Ada.

Three-quarters of an hour later, we were back on the Harley, heading out into the desert towards Big Kitty's Live Action Ranch. It was good to be out on the open road. Life was suddenly full of freedom and excitement and – at least for the time being – no one was trying to kill me. I could get used to this. Every now and then I snuck a glance behind me to see if anyone was following us, but for once we were ahead of the game. All we had to do was persuade one of Big Kitty's keepers to take a day off sick and lend a hand with keeping Larry the Liger under control. Then Ada and I could get into the operations room, gather whatever evidence we needed and be back home in time for tea.

Big Kitty's Live Action Ranch turned out to be something of a disappointment, however. Just before the turn-off from the highway a vast hoarding promised us 'over a hundred tons of crazy carnivorous critter action', but the paint was peeling and a crudely written note plastered underneath informed us that it was closed until further notice. Ada took the turning anyway and, half a mile further on, we arrived at a compound surrounded by a chain-link fence. The gate was swinging aimlessly back and forth, so we got off the bike and walked carefully through, looking from side to side as we did so.

'Hello?' I called out. 'Anybody there?'

Nobody responded.

To our left, there was the remains of what I took to be the gift shop, although there was nothing left to buy there. To our right, there was a clapboard shack with the words 'Private – Staff Only' in red paint on the door. Ahead of us, the track stretched out into the distance, with caged enclosures on either side. We carried on walking, peering nervously into the first of the enclosures to make sure it wasn't occupied.

'Do you get the feeling we're being watched?' said Ada.

'Yes,' I said. 'But what by?'

'I'm kind of hoping human.'

'Me too.' If the humans had buggered off and left the animals to it, they'd be pretty ravenous by now.

There was a loud crash behind us as the door to the shack was thrown open. We both spun round to see a long-haired guy with a straggly beard pointing a shotgun at us. He was wearing a dressing-gown over pyjamas with a ragged pair of slippers on his feet.

'Stop right there,' he said. There was no great urgency in his voice, but the presence of the shotgun more than made up for it.

'It's OK, we're not going anywhere,' said Ada.

'Very happy to stand here as long as you like,' I added.

The long-haired guy did a double take. 'You British?' he said.

'Um, yes,' we both said in unison.

'I *lurve* the Brits,' said the man, raising both arms into the air. Then he lowered his gun and beckoned us over. We glanced at each other, shrugged and decided to comply.

The door to the shack opened onto a small hallway that led to a living area that gave off a definite ambience of 'inner refuge two and a half months after a nuclear attack'. Unwashed clothes were scattered over every available surface. Plates with the unscraped crusty residue of a full week's meals were stacked up at one end of a table next to a laptop with several keys missing. The floor was crunchy underfoot, particularly so in the area surrounding the overflowing wastepaper basket. In the corner there was a mattress that advertised a wide range of stains from the human bodily fluid catalogue, along with some that appeared animal in origin. The unusually broad spectrum of smells would have sent a bloodhound into overload.

The man ushered us in, then broke the gun open and removed the two cartridges before placing it on the table. Then he identified a couple of chairs underneath all the clutter and cleared them of junk with a deft sweep of his arm. We sat down, trying to avoid breathing more than was absolutely necessary.

'Coffee?' he said.

'Uh... I'm good, thanks,' I said, eyeing up the array of chipped mugs on the table and wondering what wild and heady alien cultures might be festering in them.

'Yeah, me too,' said Ada.

The man found another chair and sat down next to the table. Then he grabbed one of the mugs and reached for a bottle with no label that was half full of a colourless liquid. He poured himself a generous slug, drained it in one go and pulled a face as if it had burnt through several of his internal organs. Given the state of the bottle it came from, there was a decent chance that it had.

'Want some?' he said, offering the bottle to me.

'I think I'll pass on that as well,' I said. 'Sorry to be a party pooper and all that.' I was also thinking that I'd like to retain my eyesight for a little longer if possible.

'Your loss, man.' He leaned back in his chair and belched. 'Waldo,' he said by way of introduction.

'Pleased to meet you, Waldo,' said Ada. 'Tom,' she added, pointing to me, 'and Ada.'

'Pleased to meet you, Tom and Ada. Welcome to Big Kitty's Live Action Ranch, or at least what's left of it.'

'Thank you,' said Ada. 'So what happened?'

'Closed us down, that's what happened,' said Waldo. 'Big Kitty's in jail now, along with most of the rest of the management team.'

'Good heavens,' said Ada. 'What for?'

'Trafficking in endangered species, that's what. Lot of rich folk fancy getting their paws on a cute little old big cat cub. Been going on for a long time. Would have gone on forever if someone hadn't snitched on them.'

'And you were working for them?' said Ada.

'Oh yes,' said Waldo, breezily. 'I did all the footwork, man. Smuggling the little critters in and out of the big mansions all over the country.'

'So how come you didn't get sent to jail?'

Waldo looked up at the ceiling and coughed. 'Well, I guess I was the one who did the snitching.'

'Ah,' I said. 'Hence the shotgun.'

'Nah, I just like guns, man.' He leaned forward and met us both in the eye for the first time. It wasn't easy at first to tell how old he was. He could have been a regular guy in his fifties and, then again, he could easily have been a couple of decades younger but on the far side of the kind of life where the years counted double. 'So what brings you to Big Kitty's?' he said.

'We've got a problem,' began Ada.

'With a liger,' I added.

Waldo whistled. 'Liger, eh? Man, they're big bastards, ligers. That is the big cat of choice for your real macho man.' He paused and then narrowed his eyes. 'Don't see many of them either. You sure it's a liger?'

'Oh yes,' I said. 'They told me it was one when I escaped from it.'

'Uh-huh,' said Waldo. 'I know a little about ligers. We bred a couple here – sold one for a goddamn fortune and kept the other for a time until it – um – got acquired by a third party.'

I glanced at Ada.

'Who was the third party?' she said. She was trying very hard to contain her excitement.

Waldo took a deep breath. 'Casino owner by the name of Robert J Merritt III,' he said. 'Owns the Gran Pelícano over in Vegas.'

'We know,' I said.

Waldo raised an eyebrow and poured himself another slug of the anonymous firewater. 'Is this story of yours going where I think it might be going?' he said.

'It could be,' said Ada. 'The thing is, there's a thing we need to get to that is in the same place that they keep this liger.'

'Just to be sure, this is Mr Merritt's liger that we're talking about here?'

'Yep,' I said.

'Well, couldn't you just politely ask Mr Merritt to move his little liger out of the way?' said Waldo.

'Well, that's the problem,' said Ada. 'We don't really want to bother Mr Merritt.'

'He's a very busy man,' I added.

'Extremely busy,' said Ada.

'Hmm,' said Waldo. 'I could be wrong here, but I'm beginning to think that you're thinking that maybe someone who may have, uh, bonded with this little kitty of Mr Merritt's in the past may be able to keep her – it is a her, by the way – from bothering you while you do whatever you have to do to this thing that you need to get to?'

'That's basically the gist of it,' said Ada, 'yes.'

'And we do this after we have somehow broken into Mr Merritt's big old penthouse suite?'

'Indeed we do,' I said.

'And you have a plan for this?' said Waldo.

'We're working on it,' said Ada.

Waldo steepled his fingers and closed his eyes. 'First of all,' he said eventually, 'you should know that ligers are extremely dangerous animals.'

'I know that already,' I said. 'As I said, I've actually met this one. I only just escaped.'

Waldo gave me a look of grudging admiration. 'Well, you should also know that Robert J Merritt III is an extremely dangerous human being.'

'We know that as well,' I said. 'One of his men has tried to kill me at least once already.'

Waldo gave us a big grin. 'Well, then. The other thing you gotta know is that I hate the fucking bastard.'

Ada's eyes lit up. 'Does that mean you're going to help us?' she said.

'Too goddamn right it does,' said Waldo. 'Where do I sign?'

Chapter 11

'Do you think he's safe?' I said to Ada as we walked back to the entrance.

'No,' said Ada, 'but I can't see any alternatives right now. If we're going to get access to Merritt's system, we need to neutralise the gatekeeper.'

'I guess so.'

'Ah, it'll be fine, Tom. Have I let you down yet?'

This was a fair question. Ada had already saved my life twice after all.

'Come on,' she said. 'Let's get back to Vegas, grab some lunch and then find your friend who can get us into the big man's lair.' We climbed onto the Harley and Ada gunned the engine.

Back in Vegas, we stopped off at the Heartbreak Motel to check if anyone had been in my room again.

'Interesting way to live,' said Ada, giving the chaos of the place the once-over. 'Waldo would feel very much at home.'

'That's not fair. It's been ransacked.'

'Again?'

'Well, no, I left it in a ransacked state. Didn't seem worth tidying up.'

'Has anything been moved since you were last here?'

'Nope. Not as far as I can see.'

'Well, that's something. Still, you probably shouldn't stay here any longer. How was the floor last night?'

'I've slept on worse.'

'Well, I think that answers the next question.'

I quickly gathered up all my things and threw them into my bag. Then I checked out and we found a bite to eat at a branch of Subway just down the road from the motel.

'How many scams, cons and other rackets are there going on in this city, do you think?' said Ada, leaning back against the wall next to Subway and taking a bite out of her falafel sandwich.

'I think it'd be easier to list the things that weren't scams, cons or rackets in Vegas,' I said.

'Pretty poor view of humanity you've got there, Tom.'

'It's realistic. Also, it's nice to be pleasantly surprised when I'm sometimes proved wrong.'

'Do you think we can pull this heist off?'

'I think our chances are roughly one in a hundred right now.'

'And you're happy to bet on those odds?'

'I've had worse odds against me recently, to be honest.'

'Really?' said Ada, tilting her shades forward to reveal questioning eyes. 'Is there something you'd like to share with me?'

'Not now,' I said.

'Well, all I'm going to say is that if you're cool with hundred to one against you, I'd stay clear of Mr Merritt's tables for the foreseeable.'

'You seem cool with it as well, though.'

'Maybe I have a more optimistic view of our odds.'

'Or maybe you're just reckless.'

'Maybe I am. I've always tended to live life within touching distance of the edge. Finished?' I nodded. She dabbed her mouth with a tissue, threw the packaging in the bin and brushed her hands together to get rid of the crumbs. Then we got back on the bike and headed off downtown, where we left the bike and my stuff at El Gran Pelícano.

I left Ada drawing up plans for the break-in and I headed over to the rear of the convention centre. All I had to do now was keep clear of Jolene until either María or Jesús made an appearance, so I crouched down behind a large dumpster. After about ten minutes the Vegas heat began to get to me and I started to doze off. I shook myself awake again. 'Come on, Winscombe,' I said to myself. 'You have one job. Do it.'

'Not-Diego?' came a voice.

'Uh?' I said. Oh Christ, how long had I been out? It was beginning to get dark.

'Not-Diego, it *is* you!' said María.

'Oh, thank god it's you,' I said.

'Too right. If it had been Jolene, baby, you wouldn't have woken up ever again.'

'What have they been saying about me?'

'Well, for a start, you're a dirty little sneak and you ain't who you say you are.'

'I don't think I ever did say who I was,' I said. 'All I ever said was that I wasn't Diego.'

'This is very true. Also they said that if anyone finds you, there's a five-hundred-dollar reward for handing you over to Don.'

'Ah.'

'But don't worry, the bastards just fired me for no good reason, so maybe I'm not going help them out this time.'

'Well, thank god for that.'

'Then again, perhaps if I hand you over, they'll give me my job back? Plus the five hundred dollars? It's kinda tempting.'

I stared at her for a few seconds.

'Well?' I said. 'Are you going to? Just so I know if I'm going to have to make a run for it.'

She considered this. 'Nah, we're good,' she said eventually. 'So how come the dog returned to his vomit?'

'I need to talk to you.'

'Well, that's good because here I am.'

'No, not here. Not safe. Can you come up to my room in the Gran Pelícano?'

María gave me an odd look. 'Jeez, are you Brits always like this? I'm trying to work out if it's cute or just weird.'

'Oh god, no, it's not like that.'

María frowned. 'And now I'm wondering,' she said, 'whether to feel like I've just been insulted.'

'No, no, no,' I said. 'Look, if it helps, my friend, Ada, will be there too.'

'Ah, now it's definitely getting weird.'

'Oh god,' I said, 'you're doing this deliberately, aren't you?' I looked at her, but her face remained inscrutable. 'Look, I'm going anyway. Come with me if you fancy doing something interesting.'

I stood up and peered over the rim of the dumpster. The coast seemed clear, so I walked off towards the front of the convention centre and then across to El Gran Pelícano. As I made my way inside the hotel, I noticed María scurrying through behind me.

'Decided to come along then?' I said.

'For the moment,' she said.

We went up to Ada's room and I introduced the two women. They both perched on the end of the bed while I pulled up a chair opposite them.

'So what's the deal?' said María.

'How much has Tom told you?' said Ada.

'Well, the first thing he *hasn't* told me,' she said, 'is that he's called Tom, so maybe we should start from there.' She turned to me. 'Hi, Tom,' she added, holding out her hand.

'Yeah, sorry about that,' I said, shaking it.

'Nice to meet you,' said María. 'Anything else I need to know? Like you're working for the CIA or something?'

'No, we're entirely independent of any governmental agency,' said Ada.

'You sure about that?' said María. The question was directed at me this time.

'Definitely not,' I said.

''Cos as – ahem – Tom knows, my status is a little fluid.'

'No, we're definitely not government,' said Ada. 'I just want to get my bike to work and Tom wants his alpacas back.'

María stared at us for a moment, started to speak and then gave up. Then she tried again, but once more the words didn't quite come.

'Don't sweat about the details,' I said. 'The first question is: am I right in thinking you've worked for Merritt for a while?'

María counted off on her fingers. 'Five years now.' She sighed. 'It was OK while it lasted.'

'So I take it you've been up to the penthouse here a few times?' said Ada, pointing towards the ceiling.

'Listen, I worked as a maid at this hotel for the first eighteen months I worked for Merritt. I know this place like I know the wrinkles on my mother's face, God rest her soul.'

'Perfect' I said. 'All we need to know is how to break into the penthouse.'

'Why would you want to do that?' said María.

'Ada just explained that,' I said. 'Bikes and alpacas.'

'You're not going to find them up there,' said María. 'Although you might come across a big cat or two.'

'We've got that issue under control,' said Ada.

María raised an eyebrow. 'Uh-huh?' she said.

'Uh-huh,' I said.

'OK, well, I'm assuming you don't want to use the private lift,' said María.

'Correct,' I said. 'We'd need to take out the front desk first if we did that. And then the lift operator guy.'

'Is it just you two?' said María.

'Well, us and one other,' said Ada. 'Our big cat guy.'

'You have a big cat guy on your team?'

'Yeah,' I said. 'Like I said, we have the big cat situation under control.'

'Jeez,' said María. 'What kind of crazy are you people?'

'Very,' said Ada. 'But also ruthless and dedicated.'

María thought for a moment and then stood up and walked over to the window. She looked down towards the ground and then up towards the sky. 'It's a long way down,' she said. 'How are you with heights?'

'What do you mean?' I said. I wasn't sure I liked the implication of the question.

María came back and sat down again. 'OK,' she said. 'From what I can remember, there's an emergency stairway that goes all the way up to the top of the building. But the problem we have is that, on the top floor, the entry point is a doorway that only opens from the inside.'

'Shit,' said Ada.

'However,' said María, 'there is another way in. Via the balcony.'

Ada and I glanced at each other. I didn't like the sound of this and I got a strong impression that she didn't either.

'Did you say balcony?' I said.

'Yep,' María confirmed. 'Big glass-sided thing. Was going to be a fancy, high-level swimming pool but then they found out it couldn't take the weight of the water. Got a piece of paper I can draw on?'

Ada located my pad of paper and pen from the previous night and handed them to María. The sequence of images that she scribbled was no less alarming than the vision that had already been implanted in my brain.

'Oh,' I said.

'Right,' said Ada.

'Yeah,' said María. 'You guys OK with heights?'

Ada and I exchanged looks. 'Not great,' I said. I didn't think this was the appropriate time to mention the fact that I'd once launched myself out of the top floor of a burning building in Minsk with an emergency executive parachute strapped to my back. I didn't want the fact that I had done this once and somehow survived the experience to serve as any kind of indication that I was in any way happy about working at any level beyond two feet or so above ground.

'Yeah, well,' said María, with a shrug. 'It's the only way you're going to get in there.'

'OK,' said Ada. 'So let's break this down.' She pointed to the first image in the sequence, which featured a rectangle sticking out of the side of a building. There was a stick man apparently leaning out of the window beneath it, dangling a rope. 'You're suggesting,' continued Ada, 'that someone leans out of the window of the room on the floor below and somehow attaches a rope to the side of the balcony?'

'That's about it,' said María.

The second image featured the stick man shinning up the rope. The rope was now dangling alarmingly in space.

'And then they all climb up the rope onto the balcony above and into the penthouse,' I said.

'Yep,' said María. 'Although strictly speaking, you only need one of you to do the balcony shenanigans and they can let the rest in through the stairway.'

'How do we attach the rope to the balcony?' said Ada.

'Could use some kind of a grappling hook,' I said.

There was a long silence. Then Ada turned to me.

'That's a really good idea,' she said.

'I was thinking we could get Waldo to do that bit,' I said.

'Who's Waldo?' said María.

'He's the big cat guy,' I said. 'Interesting character.' I turned to Ada. 'Well?' I said.

Ada's reply was non-verbal and centred almost entirely on her eyes which, despite being covered by her usual shades, somehow managed to convey her unshakeable impression that (a) Waldo was most definitely not the person for this particular job and (b) neither was she.

'Shit,' I said. Why hadn't I kept my mouth shut?

There was a long silence as we all contemplated what was being proposed. I maintained the silence for a little longer, as I had a bit more contemplating to do than the other two people in the room.

'Two questions,' I said eventually. 'First one. How do we get access to the room below? And please don't tell me we get there from the room below that, because I may just run screaming from here and never come back again.'

'We steal the key from reception,' said María. 'Room 3204.'

'Cool,' said Ada, glancing at me with a smile. She'd noticed the use of the word 'we' as well. We had María well and truly hooked. 'You're on board with this too?'

'What?' said María. 'If I'm joining your gang, I'm not coming in as a part-time member. It's not just about the planning. You're gonna to need a guide once we get up there, too. Believe me, I've been there a few times and I used to get lost regularly.'

'OK,' I said. 'But you're sure you want to be part of this?'

María shrugged. 'Sounds fun,' she said. 'And it's not as if I'm busy right now.'

'OK,' I said, 'Let's go back to the balcony. What if there's someone in the room below?'

'We take them out,' she said.

'Hold on a minute,' I said. 'You can't just start killing people.'

'Nah, I didn't mean literally taking them out,' said María. 'Just tie them up a bit.'

'Right,' said Ada, exhaling.

'Where do we get a grappling hook from?' I said.

Ada was tapping away at her phone. 'Ordered one from Amazon Prime. Arriving tomorrow. Got a rope ladder too. Bit more comfortable than just a bit of rope.'

I frowned at her. 'Isn't it a bit risky to have a grappling hook delivered to the very hotel where you're intending to use it?' I said.

'Thought of that,' said Ada. 'There's a drop-off point just down the street.'

I had to admire her level of planning and I was beginning to feel that this might actually work. There was the tiny detail that I was going to have to climb a rope ladder suspended several hundred feet up in the air, but I would have to cross that bridge when I came to it. We were even building a proper multi-disciplinary team: Ada with her organisational skills, Waldo with his understanding of big cats and María with her insider knowledge of Merritt's operation. Plus myself, of course.

'So what's the schedule?' said María.

'Good question,' I said. This was something we hadn't really got round to discussing.

'I think we should move as soon as possible,' said Ada. 'Don't forget Don and Jolene are still out looking for you, Tom. We need you to climb that ladder, so we don't want you to get caught.'

'Is that all I'm good for?' I said.

'It's a very important role,' said Ada. 'I knew we'd met for a reason.'

'The online gambling convention opens the day after tomorrow,' said María. 'There's been a few odd rumours floating around.'

'What kind of rumours?' I said.

'Big poker game taking place around then,' said María. 'Ultra high stakes.'

'Ooh,' said Ada. 'Would I be right in thinking that your ex-boss might be taking part?'

'That's the word on the street,' said María.

'So that would ensure that he was out of the way, right?' I said.

'Yep,' said María. 'Along with his entourage.'

'Perfect,' said Ada. 'Can you find out any more?'

'I'll try,' said María. 'I'm going out for a drink with my old co-workers this evening, so I'll see what they're saying.'

'Great. So that's agreed then. We coordinate the raid on the penthouse with the poker game. As soon as they get down to business, Tom breaks into room 3204 while Waldo, María and myself go up the stairway and wait outside the emergency escape door. Tom slings the grappling hook over the edge of the upstairs balcony, climbs up the rope ladder and sneaks into the penthouse. He lets the three of us in and together we all find our way to the jungle room. Waldo keeps the Liger happy and María guards the door while Tom and I find the main computer system.'

'What are you going to do then?' said María.

'Oh, I'm sure we'll work out something,' I said. 'It'll probably be obvious what's going on.'

'Right,' said María. 'So we don't have a computer expert on the team, then?'

'Not as such,' said Ada.

'Nope,' I said.

María looked at us both and let out a deep sigh. 'Great,' she said. 'We go to all that trouble and you don't have anyone who understands computers. Nice work, guys.'

'I'm sure we can work something out,' I said. 'Can't be that hard. All we need to do is prove that it's the source of a massive ransomware scheme. I've got people I can call if we need help anyway. They'll be keen to get me out of there and home safely because I'm going to be their sperm donor.'

Ada and María both stared at me for several seconds.

'What?' I said. 'It's not the first time anyone's done that.'

'No, no,' said Ada. 'I'm sure it isn't. It's just you haven't mentioned this before.' She was looking at me with her head on one side.

'Maybe you need to take a sample and post it to them anyway,' said María, gravely. 'Just in case. Let us know if you want us to turn the other way.'

'Wish I'd never mentioned it,' I said.

'No, I'm glad you did,' said Ada, suppressing a smile. 'Reveals a whole new side to you.'

'Yeah well, I have all sorts of hidden depths,' I said. I had a strong impression that both women were laughing at me and I didn't like it. However, the digression had served one purpose in that María was no longer thinking about what we were going to do when we got inside and that at least gave us a bit more time to come up with something a bit more concrete. If only Dorothy was here and not halfway across the planet in Macau. She would have been perfect for the job.

It was getting dark outside and María needed to leave. We agreed to keep in touch and let each other know if we had any more bright ideas. Ada and I rode down in the lift with her to make sure she got clear of the hotel safely.

'Do you really think this is going to work?' I said to Ada as we turned to go back up to her room.

Ada was silent for a few moments.

'Well,' she said eventually. 'There's only one way we're going to find out.'

Chapter 12

Ada's floor wasn't any more comfortable the next night but I managed to get something close to the right amount of sleep anyway. Once I was awake, I dragged myself towards a few degrees shy of vertical and went to the bathroom. At least I didn't have a raging hangover, so that was one positive thing to add to the list straight away. Not only that, but I now had all my things with me, so I managed to make myself feel human again, which was a good baseline to start building on.

As I came back into the room, I heard a 'ping' from Ada's phone.

'OK, Tom,' she said, from where she was standing by the window, 'action stations. María says the big game is on tonight. Merritt's staff have been given the night off so we should have the place upstairs to ourselves.'

'Tonight?' Crap. I'd planned to have at least another full day of existence before I ended up plummeting out of the sky followed closely by a poorly aimed grappling hook.

'Yep. Shit just got real.'

'Better make sure Waldo's on board.'

'I'm just messaging him now. Although I'm guessing he's not an early riser.'

'What time does it all kick off?'

'Nine p.m.'

'Well, that gives us time to organise.'

Ada was preoccupied on her phone for another minute or so, then she announced, 'Aha! Guess what, Tom. Your grappling hook's arrived.'

'Awesome,' I said.

'It's a great thing you're going to be doing,' said Ada. 'And the really good thing is that it's going to be dark, so you won't be able to see the ground.'

'Until I'm just about to land on it,' I said.

'Nonsense. You're good at this stuff. I can tell.'

My phone pinged. It was Ali.

Hey, Fuckwit, it said. *Hope you haven't forgotten.*

Forgotten what, I thought? Whatever it was, I certainly had.

Remind me again, I texted back.

You have forgotten, haven't you? came the response. *It's Clinic Day. We're here waiting for you to make a deposit into the bank of Patti.*

Oh shit. It was today. I had forgotten.

Look, something's come up, I said. God, that was weak.

There was a very long pause before anything came through.

What do you mean, something's come up?

I'm somewhere else, I said. *Quite a long way away. Really quite a long way away altogether. I'm in Las Vegas.*

Fucking hell, Winscombe.

Sorry.

You fucking will be.

Ada was looking at me. 'Trouble?' she said.

'Ah, nothing,' I said. Then my phone pinged again.

Also, Patti wants to know what the fuck you're doing in Las Vegas? said Ali. *And she wishes it to be known that those are her exact words.*

This was serious. I'd never known Patrice to swear before.

Look, I'm really sorry. I'll be back in a couple of days, I said. *Just need to sort out a few things.*

Oh sure. Do you know anything about biology? came the response.

'Doesn't sound like nothing,' said Ada.

'Well, there's very little I can do about it now,' I said, willing my phone to stop beeping at me.

'Sure?' Ada looked concerned, although I felt it probably wasn't just for my well-being. It was more to do with making sure that nothing got in the way of the current project.

'No, it's fine. I just let someone down, that's all.'

'That's not quite as reassuring as I'd hoped,' said Ada.

I could see her point.

'It's OK,' I said. 'You can rely on me.' I made a point of not saying what it was that she could rely on me to do, just in case what it was turned out to be 'cock the whole thing up'. The light of dawn in Vegas isn't particularly cold, but it was still having a pretty good go at highlighting every single tiny flaw in the plans we'd drawn up in the preceding twenty-four hours, and my confidence was ebbing away towards rock bottom.

'We need some breakfast,' said Ada. 'And then we'll sneak out and get the grappling hook.'

'Cool.' With any luck, something would have gone wrong with the order and they'd have sent something completely different. Grappling Hook's difficult third album, for example. You know, the one where they brought in the Tuvan throat singers? Nah, preferred their earlier work, mate.

Unfortunately it turned out that the grappling hook, along with the rope ladder, were exactly the ones that Ada had ordered. Holding them in my hand made the whole ridiculous prospective escapade suddenly seem a lot more concrete.

'Any good at knots?' said Ada.

'Nope,' I said. 'I can just about manage to tie my laces on a good day, but that's about it.'

'Well, we've got all day to practise and a YouTube full of willing demonstrators I'm sure.' She produced her laptop from her bag and thrust it at me where I sat on the bed. I hesitated. 'Go on, then,' she said, clapping her hands. 'Get cracking!'

I held up the top of the rope ladder and let it trail on the ground. Was this really going to support my weight while I climbed up to the balcony?

'Are you sure there's no other way of doing this? Couldn't we just barge the door to the stairway open?'

'We'd probably set off some kind of alarm.'

'We might not.'

'Well, we might not,' said Ada, 'but we can't rely on that.' She began to rummage in her rucksack. After a few seconds she produced a roll of gaffer tape, a number of metal clips and some thinner nylon rope. She tossed them over to me and I duly caught the gaffer tape and dropped the rest.

'What's all that for?' I said.

'Gaffer tape's always useful,' she said. 'And you can use these carabiner things and rope to rig up a safety harness. Basically you'll be able to attach yourself to the ladder at each stage as you go up. Every time you move a rung up, you just detach the carabiner and re-attach it at the next level up. You'll need to be good at your knots, though.'

'Great.'

'So get going!'

'Yes, miss.'

'Let me know when you're done.'

She slouched down in the chair opposite and began to read a book, as if she hadn't another care in the world. I had a quick hunt around YouTube and eventually settled on an enthusiastic but slightly alarming redneck with a resplendent beard and some unsettling tattoos who went under the moniker of Nick 'the Knot' Knotty. I sat transfixed for the rest of the morning, tying and untying the nylon rope to the carabiners and the rope ladder to the grappling hook. By midday, I was beginning to feel I was getting the idea of this. I was pretty sure that I should be using a bosun's double cloverleaf lock hitch to attach the rope ladder to the grappling hook, although a quick check in the comments revealed that there was a substantial portion of the online community that felt that Nick 'the Knot' Knotty had made this one up just to show off. Apparently, this was something he was well known for. Well, I liked the knot anyway and I was prepared to trust my life to it.

Ada came over and picked up my knot. 'What's this?' she said.

'It's a bosun's double cloverleaf lock hitch,' I said.

She gave it a gentle tug and the entire thing unravelled. 'Neat,' she said, tossing the rope back to me.

'Oh,' I said. Maybe I wouldn't trust my life to the bosun's double cloverleaf lock hitch after all.

'Maybe just stick to something more straightforward for now, Tom.'

I went back to my tutorials.

At around three o'clock in the afternoon, María arrived, brandishing an electronic room key.

'Ta da!' she said. 'Room 3204 is ours!'

'Cool,' said Ada. 'Anyone staying there at the moment?'

'Nope,' said María. 'For some reason, they don't use it unless they're completely full.'

'Maybe Merritt likes to keep some clear space between him and his guests,' I said.

'Well, if that's the case,' said Ada. 'I think we should go and take a look. Coming, Tom?'

'Definitely,' I said. I had seen quite enough of Nick 'the Knot' Knotty.

Room 3204 was shrouded in darkness. We went over to the window and threw back the curtains. I looked down and my stomach detached itself from its moorings. I tried to reassure myself that it didn't actually make a lot of difference to the end result whether I fell from thirty feet or three hundred, but this didn't help at all.

'It's a long way down,' I said.

'But only a short way up,' said Ada, pointing to the balcony that loomed above us. 'Think of it that way.'

I began to wonder about how I was going to swing the grappling hook in order to get it to engage with the balcony and a whole list of problems began to present themselves to me, with one very big one right at the top.

'OK,' I said. 'This window here. It doesn't open.'

'Well, obviously,' said Ada. 'You don't want your guests falling out, do you?'

'But that's a bit of a stumbling block for our plan, isn't it?' I said.

Ada reached into her rucksack and produced what looked like a wooden handle with a suction cup attached to each end. She waggled this at me by way of explanation.

'What's that supposed to do?' I said.

'Simple,' said Ada. 'We attach this to the window, then score round the edge of the glass with a knife. Then we remove the glass.'

'It's double glazed.'

'Then we do the same thing twice,' said Ada, with a hint of exasperation. 'God, Tom, are you always this negative?'

The silence that followed was broken by a barely perceptible click coming from the door.

'Shit,' hissed María. 'There's someone coming! We need to hide!'

All three of us simultaneously realised that there was in fact nowhere to hide in the room and that the only thing to do was to flatten ourselves against the window, draw the curtains in front of us and hope for the best. It was probably just someone from the hotel checking that everything was OK and that nothing needed cleaning.

'Oh, babe,' said a male voice. It was a familiar voice, although I couldn't place it at first.

'Been a tough few days,' said a second, female, voice, slightly muffled. I recognised this one too, but I still couldn't place it. Neither of them said anything more for a while, although there was definitely a significant amount of non-verbal communication going between the two of them, along with the rustling of clothes.

'What if he's still running round?' said the man. Bloody hell! It was Don. 'He could wreck everything, babe.'

'Try not to think about it, Donnyhoney,' said Jolene.

So this was why everyone kept Room 3204 clear. It was the senior staff shag-suite. I glanced at Ada and María, both of whom were displaying body language indicative of a similar amount of internal cringe to that which I was currently experiencing. I also had the feeling that this was about to get much, much worse.

'I can't get him out of my mind,' said Don. 'How can a little runt like that get away from me? Twice!'

'Sssh, babe,' said Jolene. This was followed by more rustling and then the sound of a fly being unzipped, after which Jolene appeared to be lost for words for quite some time. Either that, or she had simply been brought up not to speak with her mouth full. 'You all right, hun?' she said eventually.

'Just not happening today, babe,' said Don.

'Let's go and lie down, Donnyhoney,' said Jolene. This was followed by a prolonged sequence of clothes rustling and then a couple of tired creaks from the bed as if it had seen all this before and knew what was coming next.

'Bloody hell,' I muttered under my breath. Ada kicked my shin by way of response.

There was now a sustained period of moaning and creaking coming from the direction of the bed, broken by a sudden cry of 'What's that?' from Don.

'What's what, honey?' said Jolene.

'There's a bag on the floor over there,' said Don.

'Don't worry about it,' said Jolene. 'Try and stay focused.' I sensed that Jolene was getting frustrated with Don. But Don was having none of it. I heard him get out of bed and pad over to where Ada had left the bag, right next to our position hiding behind the curtain.

'Yes, it's definitely a bag,' he said. 'Might be a bomb.'

'Oh for god's sake, Don, leave it and come back to bed,' said Jolene.

'Can't see what's in it,' said Don. 'Looks like tools or something.'

'Don!' said Jolene.

'Something with a wooden handle—'

At this point, Ada threw back the curtain and the three of us were briefly treated to the sight of the twin hairy globes of Don's naked buttocks, bending over Ada's rucksack. Then Ada pushed past him, sending him sprawling, before leaning down to swipe her bag out from under him. 'I'll have that,' she said to the speechless and prone Don as we all filed past him towards the door, waving at Jolene as we did so. She responded by pulling the duvet up to her neck.

Ada slammed the door shut. Outside in the corridor, she issued instructions.

'OK, we need to split up,' she said. 'Tom, take the lift. We'll go by the stairs. Reconvene back in the room in half an hour, OK?'

'OK,' I said. The lifts were right at the other end of the corridor, so on reflection I would have preferred to take my chances on the stairway with the others. But I appreciated that it was better tactically to confuse them by going in opposite directions, so I began to run down the corridor. When I reached the lifts, both of them were right down on the ground floor. I stabbed at the call button and waited for one of them to start moving. After an age, the lights above the one on the right indicated that it was beginning to lurch upwards.

There was a clatter behind me. I turned round to see Don staggering out of Room 3204, one shoe on and one shoe off, trousers round his ankles, attempting to dress himself as he went. This was never going to end well and, sure enough, half a dozen steps on, he tripped over and fell flat on his face. As he struggled to get up again, Jolene emerged from the room, pulling on her blouse as she did so.

'Which way, Donny?' she said.

Don, now on his knees, twisted round and waved vaguely in the direction of the stairwell. 'Two of them must have gone that way, babe. I'll take this guy.' He pulled himself to his feet, hiking at his trousers and lurched forward before falling over again. I turned back to the lift and realised to my horror that it had stopped at the fifteenth floor.

'Come on, come on, come on,' I muttered to myself. It was only a matter of time before Don finally got his act together and caught up with me. Meanwhile, the lift on the left had sparked into life and was also heading my way. The reason for this soon became obvious as the right-hand one was now heading back downwards. My fate was now basically in the hands of a lift algorithm.

Meanwhile, Don had sensibly decided to take a few seconds to sort out his trousers and shoes before resuming the chase. I glanced up at the lights again. The lift had picked up speed and was now at floor eighteen. Then suddenly it was at twenty, twenty-two, twenty-four, six, eight, thirty, PING! The doors opened and I threw myself in, narrowly avoiding an elegant lady in her seventies trailing a suitcase on wheels. If looks could kill, the one she gave me as she exited would not only have done for me but would have probably set fire to the lift as well.

As I pummelled the button to close the doors and go down to the ground floor, I saw Don hurtling, out of control, towards me on a course that was perfectly calibrated to collide with the lady with the suitcase. Which he duly did, just in time for me to wave goodbye to him as the doors finally joined together and the lift began its descent.

The downward journey gave me just sufficient time to recover my composure and get my breathing back to somewhere close to normal. As the doors opened on the ground floor, I marched resolutely out into the lobby and then out into the heat of the Las Vegas afternoon.

What now? Presumably Don, having failed to nab me on the thirty-second floor, would also be heading downwards just as soon as the right lift arrived. Jolene would be chasing Ada and María down the stairs, although they had a pretty good head start on her and I felt there was a good chance of them avoiding any trouble with her. The fact that Don and Jolene were up there for the purposes of conducting some kind of illicit liaison also played into our hands, in that they were unlikely to call for reinforcements in case they had to concoct a story as to how they'd come across us. It was just the two of them against the three of us.

I headed off downtown with the intention of merging in with the crowd. From time to time I risked a backward glance to check that no one was following me. There wasn't. Half a dozen blocks down from El Gran Pelícano, I ducked into a bar and ordered myself a beer. I sat at the counter on a bar stool listening to the conversations going on around me and tried to imagine what it would be like to live an ordinary life.

'Gonna be quite a game tonight,' one of the old guys at the bar was saying. 'Old Merritt don't show up unless the stakes are sky high.' He was wearing a lumberjack shirt over a pair of jeans and cowboy boots. He had a finely sculpted grey moustache and goatee beard.

'He sure don't,' agreed the guy on his right, who was completely bald apart from a thick black moustache that had almost certainly been dyed that colour. 'But when he does, can get pretty wild. I remember back in July '17...'

''18,' said the first guy.

'You sure, Bub?'

'I would stake my goddamn life on it.'

'Well, anyways, July '18 then...'

'It was August.'

'I don't care when the hell it was, man! Do you remember the game?'

'Nah, I was outta town.'

There was a long silence.

'So who else is in?' said the first guy eventually.

The second guy began counting off on his fingers. 'Well, there's Novak.'

'There's always Novak.'

'And Wild Cat Willy Dunoon, of course.'

'Didn't he win in '18?'

'Gotta be kidding, man. As if Merritt's ever gonna allow anyone else to win.'

'Well, he nearly won.'

'Waste of time coming second, Bub.'

The first guy considered this and then nodded slowly. 'What about the new guy? Jimmy something?'

'Don't know diddly squat about him. Not worth knowing about either. If we've never heard of him, he ain't gonna be no good, that's what I say. Not worth a bet anyways.'

There was another long silence.

'So who you puttin' your money on, Bub?'

'Reckon I might put a few dollars on Merritt.'

'Yep. Merritt's always worth a few dollars.'

There was another long silence. I looked at my watch. By the time I got back to El Gran Pelícano, the half-hour would be up. Much as I was enjoying the repartee of my two companions at the bar, it was time to go. With any luck, Don would have given up and gone home by now, so it should be safe to venture out again. Also, I still hadn't finished my knotting. I texted Ada.

You back yet? I said.

The reply was almost instantaneous: *Been back here for twenty-five minutes. Can you grab a few pizzas on the way? Starving.*

Chapter 13

Ten minutes later I rejoined the gang in Ada's room with a selection of takeaway pizzas that were long on quantity even if they fell short on quality. I offered a slice each to Ada and María and they recounted how they had managed to give Jolene the slip by exiting the emergency stairway halfway down, waiting for her to go past and then heading back up again.

'Are we safe here, though?' I said. 'What if they check with the guys on the front desk? You've quite a distinctive look, Ada.'

'Unlikely,' said Ada. 'Remember, they've got no idea that I'm staying here. The only one they know about is you, Tom, and they think you're back at the Heartbreak Motel.'

I thought about this. 'I'd still be happier to be out of here,' I said.

'Well, we will be soon,' said María, checking her watch. 'In fact in just under three hours' time.'

'I need more pizza,' I said. I felt suddenly very anxious.

At that moment there was a knock on the door and I immediately felt even more anxious. María and I glanced at each other. I briefly wondered where I could hide. I had a feeling that behind the curtains wasn't going to cut it a second time.

'Relax,' said Ada, going to the door. She opened it and Waldo strolled in. He was wearing full camo gear, up to and including full face paint, with an extremely full knapsack slung over his shoulder. He had a baseball cap on his head, which he tipped to Ada and María.

'Good evening, ladies,' he said. 'Hi,' he added, giving a perfunctory nod in my direction. Then he squatted down in the corner, rummaged in his knapsack and removed what looked like some kind of assault weapon, which he proceeded to strip down and clean with an oily piece of rag.

'This is María,' said Ada.

'Hi, María,' said Waldo, doffing his cap again.

'Pizza?' I said.

He held out a filthy hand. I placed a wedge of pepperoni in it.

'Um… That bag of yours,' I said. 'Is it full of weapons?'

He looked up at me, his mouth full of pizza, and shrugged. I took this to mean that it was.

'I mean, we're not planning to kill anyone,' I said. 'Are we?' I glanced at Ada and María to check that I hadn't missed any change to the parameters of our forthcoming operation.

'No,' said Ada, reaching for another slice of pizza. 'Tom's right. No violence. No hostages, either, in case you were wondering.'

'Aw, shucks,' said Waldo.

'And we don't kill Larry the liger either,' I said.

'Heavens no, man,' said Waldo. 'What do you take me for? Also, as I believed I mentioned, she's a lady liger.'

'Fine,' I said. 'Carry on, then.'

'Don't worry, man, I will,' said Waldo, going back to his cleaning. I was still concerned that we'd got a literal loose cannon on our team, but he was, unfortunately, a vital member of it. We had to get past the liger. I went back to my knots. After half an hour, I felt I had something that might support my weight without unravelling completely.

'What are you doing, man?' came Waldo's voice from the corner.

'Attaching this rope ladder to this grappling hook,' I said.

'Cool,' said Waldo. 'Let me have a look.'

I handed it to him. He gave it a good tug and once again the entire thing unravelled.

'You got a parachute, man?' he said, raising an eyebrow at me.

'Yeah, very funny,' I said. 'Can you do better?'

'Well,' said Waldo. 'As I matter of fact, maybe I can.' He reached into his knapsack, withdrew another unlabelled bottle containing a colourless liquid, took a long swig from it and then put it back. Then he took the end of the ladder and fiddled around with it a minute or so until the two elements were once again securely fastened to each other. He gave it a sharp pull and absolutely nothing happened. Then he chucked it back at me. 'Try that,' he said.

This was my turn to take apart someone else's handiwork, so I went at it with gusto. But however hard I heaved and strained, I could not remove the ladder from the grappling hook. Bastard. How did people do this? More to the point, why couldn't I? Still, I had to admit that this was just what I needed.

'You couldn't make me a safety harness too, could you?' I said.

Three hours later, we were ready to move. Ada had chosen a particularly striking blue wig for the occasion and, as she led the way to the lift, I felt there was nothing we couldn't achieve between us. Carrying bags full of our equipment, we went straight up to the thirty-second floor and back to Room 3204. María presented the key again and the door clicked open once more.

We closed the door behind us and left the lights off. Then we opened the curtains, half-expecting Don and Jolene to have turned the tables on us. But the place was empty. No one was expecting us to come back to the scene of our earlier crime.

Ada walked up to the windows, selected the one with the best angle to swing the grappling hook upwards from and took out the two handles from her bag. Then she attached them to the window, making sure that the suction cups were fixed nice and tight against the glass.

'Hold those two, please, Tom,' she said. I took hold of them while Ada picked a Stanley knife out of her bag and began scoring around the edges of the window with firm, straight strokes. Then she hacked away at the corners until she was satisfied that she'd got through.

'OK, pull,' she said to me. I pulled. Nothing happened.

'Bit harder,' said Ada. I pulled as hard as I could. There was a crack and a snap and then suddenly I was flat on my back with an entire pane of glass lying on top of me. It was heavier than I'd anticipated, but at least it was still in one piece.

'Good work,' said Ada. 'Now for the outer pane.'

I got to my feet and carefully propped the sheet of glass against the wall on the other side of the room. Then I removed the suction handles and took them over to the window where Ada was positioned once again. We made sure that the handles were fixed firmly in place and then I went to grab both of them.

'Think we might need some extra muscle here, Tom,' said Ada. 'Just in case.'

'Why's that?' I said.

'Because if this one falls the wrong way,' she said, 'it rather gives the game away about what we're up to. Also, there's a very good chance it would kill someone, and we don't want that to happen, do we?'

'Ah,' I said. 'Good point.'

Ada motioned to Waldo, who was, once again, passing the time by maintaining his armoury. He put down the assault weapon that he was polishing and came over. María also came and stood behind us although, as there were only two suction handles there wasn't a lot she could do apart from make encouraging noises. Ada set to work once more and in no time at all she was through.

'OK, chaps,' she said. 'Pull!' Waldo and I gave our all and this time the glass came away first time. A blast of cold air suddenly attacked the room, sending the curtains billowing.

'Nice work,' said María.

We placed the second pane of glass next to the first one and examined our handiwork. I now had the perfect location from which to swing the grappling hook up to the balcony above. All that remained for me to do was the actual swinging.

'So then,' said Ada. 'Are you good?'

'Depends on what you mean by good,' I said. 'I'm good in the sense that all physical obstructions to what I need to do have been eliminated, but in terms of actually feeling ready to do what I need to do, I am not very good at all.'

'I'll take that as a yes, then,' said Ada.

'Fair enough,' I said.

'OK, guys,' she said to María and Waldo. 'Are we ready?'

María and Waldo grunted their approval. Waldo fished in his knapsack for his bottle of firewater and offered it round. No one took him up on the offer. He took a large swig himself, put the bottle back and hefted the knapsack onto his shoulder.

'Right,' said Ada. 'We're moving up. See you on the other side of the fire door.'

'Cool,' I said. 'What if something goes wrong?'

'It won't.'

I was glad that we had someone with a solid, if entirely misplaced, sense of optimism leading us. Right now, I wished I had some of it myself.

I walked over to the window and looked down. Las Vegas seemed a very long way away indeed, although I knew that I could get there very quickly if things didn't go as planned. I thought back to the time in Minsk when I had launched myself out of the top floor of a hotel, but that was different. For one thing, I had a parachute strapped to my back, and for another the building was on fire, so there was no real alternative. What I was about to do this time was based on a conscious choice.

Come on, Winscombe. Time to kick arse.

I took the grappling hook and rope ladder out of my bag and weighed it in my right hand. I mimed swinging it upwards and to the left. OK, I said to myself. Let's do this.

I moved over to the empty space where the window had been and dangled the hook out of it. Then I gave a swing back and launched it upwards on what I thought was roughly the correct trajectory.

It missed.

The hook was now hurtling earthwards on an ever-steepening parabola.

'Shit!' I said out loud as I realised that I was just about to lose the whole thing. I flailed at the end of the ladder that was rapidly snaking towards the open gap and just succeeded in grabbing hold of the last rung before it disappeared, taking me with it. With my free arm, I grabbed one of the legs at the end of the bed and just stopped myself from being pulled out. I waited for the inevitable smash as the hook collided with the window two storeys below me.

But there was no sounding of breaking glass: just the clank of the metal hook against the wall. I was in luck. Assuming, that is, that no one in the room below me happened to be looking out of their window at the time. For a brief moment I was now in equilibrium, and I used that time to reposition my legs against the wall under the window. I gingerly took one hand off the bottom rung and grabbed the one above, pulling it in. Then I got hold of the next one, and the one after that, until finally the hook itself made an appearance. I took hold of it and pulled it back into the room.

Hoping desperately that no one had spotted the hook tumbling down the side of the hotel, I put it down and stood up again. I paced up and down the room for a minute, trying to calm myself down. Then I was ready for another go. I took a look at the bed and formulated Plan B. I decided that it was definitely not going to fit through the window without taking out the walls on either side and I assessed the chances of this happening as being relatively low. So I went ahead with Plan B, and tied the bottom end of the ladder to one of the bed's legs using a knot that I hoped was very similar to the one that Waldo had used on the other end.

I took the hook in my hand a second time. This time, I took a few swings backwards and forwards before actually launching it upwards. I watched, heart in mouth, as it soared away from me, pulling the ladder after it until, with a satisfying clunk, it snagged itself on the lip of the near side of the balcony, just as I had planned. I gave it a tug. It didn't move an inch. I gave it a tug from the left. It still didn't budge. I gave it another tug from the right. It was still firm. I re-checked the end of the ladder. It was still tied onto the bed. So even if the hook came loose, I wasn't going to tumble to earth.

I was almost ready to trust my life to it.

I leaned out of the window, took hold of the ladder and clipped the carabiner from the right-hand side of my improvised safety harness to it. Then I attached the left-hand one. I swung my right leg out of the window and connected with the ladder lower down. Finally, gripping the ladder tight between my hands, I swung my left leg over as well.

As my leg connected with the ladder, it lurched away from me and I felt myself leaning backwards at an alarming angle. I tried to haul myself back but only succeeded in converting myself into an unexpectedly effective pendulum. As I swung helplessly backwards and forwards, I noticed with some alarm that the end of the ladder was now taut against the sharp edge of the glass where Ada had used her Stanley knife and that the motion of it was making quick work of cutting through the rope.

Sure enough, after a minute or two, the rope ladder broke free of its moorings and I was now dangling free at the mercy of the hook. I had to get moving. The further up I got, the higher my centre of gravity and the smaller the radius of the pendulum. But I had to go slowly and carefully in order to avoid jolting the hook. The slightest nudge in the wrong direction could cause it to disconnect.

I unhooked my right carabiner and moved it up to the next rung. Then I did the same with the left one. Then I lifted my right leg up and flailed about trying to find where the next rung was. At this point I made the terrible mistake of looking down and I realised that it was a very long way down indeed. Come on, Winscombe. I tried again and located the correct rung for my right leg and did the same with my left. There was a horrible scraping sound from above me as the grappling hook adjusted position. But I was a few inches closer to it than I was before.

I repeated this manoeuvre a dozen or so more times until I was right at the top of the ladder, almost within grasping distance of the lip of the balcony. However, this was the point at which I realised that this was about as far as I could go. I couldn't continue on my current trajectory, as this would involve somehow clambering over the grappling hook itself, and given how precariously balanced it was, there was no way I was going to try this. The alternative was to go to one side of the hook and haul myself up over the balcony wall. Unfortunately, however hard I stretched, I couldn't get a safe enough grip on the edge. Not only that, but there were absolutely no hand or footholds whatsoever to be had on the outside of the balcony – as María had described, it was made out of solid glass with a rim of steel running around the top. I thumped my fist against the glass, but as it was the kind of glass they make suspended swimming pools out of, it wasn't going to yield to the pummelling of a puny human. I was completely stuck. There was no way upwards and no way back that didn't involve plummeting all the way down to earth. The mission had failed and I was very likely going to die.

Chapter 14

Then things took a turn for the worse, because the doors to the balcony flew open and someone walked towards me. This was it. They would just unhook the grappling iron and watch as I fell away. I'd nurtured fantasies of being Danny Ocean, but it looked as if my end was turning out to be more along the lines of Hans Grüber's final departure from the Nakatomi Plaza.

'You OK, man?' came a familiar voice.

'Waldo?' I said. 'Is that you?'

'Well, it sure ain't one of the ladies,' said Waldo. He was leaning on the rail of the balcony, smoking one of his dodgy roll-ups.

'Any chance you could give me a hand up?' I said.

'I love your English way of politely asking,' said Waldo. 'I'd be screaming "For fuck's sake, Waldo, get me fucking outta here!"'

'Well, I can do that if you like,' I said.

'Nah,' said Waldo. ''S'OK.' He put his cigarette between his lips, then reached over the edge of the balcony and took a firm hold of my shoulders. I let go of the ladder and reached up to grasp his arms as he pulled me over the wall to safety.

'Thanks,' I said. I was beginning to get tired of having everyone save me, but it was, on further reflection, good to still be alive.

'No problem, man,' said Waldo.

I unclipped myself from the ladder, removed the grappling iron from the side of the balcony and hauled in the rope. Ada popped her head around the door.

'This place is wild,' she said, keeping her voice low.

'Wait till you see it with the lights on,' I said.

'You OK?'

'Sure. What happened?'

'Door was already open.'

'Are you trying to tell me that I was risking my life for nothing?' I said.

'It's character-building, Tom,' said Ada. 'Come on.'

I left the grappling hook in a corner of the balcony and then Waldo and I followed Ada into the penthouse suite, where María was waiting in the interior section of the disused pool that had been converted into a bar area. There were no lights on so the corridor that led up and away from us was completely dark. But there was sufficient light from the moon streaming in from the balcony to see that she looked anxious.

'What's going on?' I whispered.

'Not sure,' said María. 'I'm sure I heard footsteps somewhere.'

'Footsteps?' I said. 'I thought you said that no one was going to be around tonight.'

'Yeah,' said Ada. 'That's what we *thought*. Come on. Stick to Plan A, but proceed carefully.'

Waldo reached into his knapsack and drew out a fearsome gun that looked like it could reduce a human being to their constituent atoms at a range of twenty metres.

'No, Waldo,' said Ada. 'Your job is to pacify the liger, not to kill anyone who gets in our way.'

Waldo seemed disappointed. 'Sure?' he said.

'She's sure, Waldo,' said María.

With a grunt, Waldo slung the gun back into his knapsack in one easy move and we set off feeling our way down the corridor, following María. We'd only gone a short distance when she came to an abrupt halt and we all had to take emergency action to avoid colliding into her and tripping over into a heap.

'In here, quick!' she hissed, opening a door on her right. We all bundled into the side room and María carefully closed the door behind us. Then she pressed her head to the door to hear whatever was going on outside.

'What can you hear?' I said.

'Sssh!' hissed María, flapping her hand at me.

I looked around, trying to make out what was going on in the room around us. My eyes were getting used to the dim light and I began to see that the room was filled with an endless array of vivaria on shelves, containing a vast number of large and deeply-scary, dead insects. At least I assumed they were dead – it was hard to tell in the near-darkness. But then I saw one of them – a spider that was I'm sure was at least a foot long – appear to move.

'Guys,' I murmured.

'What?' said Ada.

'Um… I think there's an insect in that tank that wants to kill us.'

Waldo immediately whipped out his pump action shotgun and aimed it vaguely around the room.

'Easy, tiger,' said Ada, placing a finger on the barrel of Waldo's weapon. 'Are you talking about that spider over there?' she added, turning to me.

'That's the one,' I said.

'First of all, it's not an insect,' she said. 'It's an arachnid.'

'I knew that really,' I said.

'And secondly,' she said, 'it'll only kill us in self-defence.'

'Are you sure?' I said. 'Have you checked? Did you ask it? Did you say, "Hey Mr Fearsome Spider, can you please confirm whether or not you eat meat and if so, do you have any plans to kill and then eat us?"'

'Quiet, you two,' said María, with her ear still pressed to the door.

'I'm just anxious to be out of here as soon as possible,' I said.

'OK,' said María. 'Well, today is your lucky day, Tom, because the coast is now officially clear.' She opened the door again and we trooped out of the room back into the dark corridor.

'Don't suppose we could put the lights on?' I said. 'Or at least use a torch?'

'Shut up, Tom,' said María.

'I was only trying to help,' I said.

'Shut UP,' she hissed.

I got a strong impression that María didn't care for me much any more. We continued on our way, with María at the front, then Ada and myself, with Waldo, shotgun in hand, bringing up the rear. We reached a point where another corridor joined us from the left and, in the dim light, I could just make out María stopping and checking both ways before taking the left turn. We fell in line and continued to shuffle along behind her. I was trying to remember the layout of the area, but I'd approached it from the opposite side, where the private lift was, so I was hopelessly confused as to which bit connected to which.

'Stop!' hissed María suddenly.

'What?' said Ada.

'They're back,' said María.

'Who?' I said. María ignored me, but instead went to open another door on our right. However, this one was locked. María muttered something in Spanish and then fumbled in her pocket for a set of keys. After a bit of fiddling around, she managed to open the door and we all went in after her. Once again, she closed the door and jammed herself right up against the other side of it, trying to make out what was going on outside.

'Why don't I just blast the fucker?' said Waldo.

'Please, Waldo,' said Ada. 'That's really not helpful.'

'It'd solve one problem,' he said.

'Yes, Waldo,' said Ada, 'and create a whole load of other ones.'

'Ah, well,' said Waldo, 'I guess you're the strategist, Ada baby.' He put the shotgun back in his knapsack.

'Anything happening out there?' I said to María.

'Sssh,' she said.

'You didn't say that to Waldo or Ada,' I said.

'Tom,' hissed Ada. 'Shut up.'

I put both hands up in defence, realising that I was in danger of being *persona non grata* with a majority of the rest of the team, and that this wasn't a good place to be, given that my position as a member of the team tonight – risking my neck by climbing up the rope ladder to let everyone in – had been rendered redundant by whoever it was that had carelessly left the emergency door open. The fact that I had needed to be rescued, yet again, somehow only made things worse.

As my eyes were once again accustoming themselves to the low level of lighting in the room, I occupied myself by looking around. At least this time there weren't any predatory spiders lurking around, although what there was turned out to be just as alarming. Running the entire length of the far wall was an enormous fish tank, from which emanated a blue-ish glow. I went over to it and put my hand against the glass. Almost immediately, something on the other side emerged from behind an artificial rock and snapped at it. I withdrew my hand immediately and ran back to the other side of the room.

'Hey, guys,' I said. 'Think this is the piranha room.'

'OK, cool,' said Ada.

'Piranhas,' I said. 'You know. Those things that strip the flesh off your body in a couple of minutes.'

'Best to stay clear of the tank then,' said Ada.

'I think I'd like to be somewhere else,' I said.

'Stop panicking,' said Ada. 'They're not going to eat you.'

'They just tried,' I said.

'Yes, but you were on the other side of the glass,' said Ada, as if she was talking to a child. I didn't like this.

'What if the glass broke?' I said.

'Then they'd probably drown,' said Ada, 'or whatever it is that fish do when they run out of water.'

'Are you telling me to shut up again?' I said.

'Basically yes,' said Ada.

'I still don't like it here,' I said.

Ada ignored me.

'At least there haven't been any snakes yet,' I added.

'I can handle snakes,' said Waldo.

'Well, that's something,' I said.

'Will you lot please SHUT THE FUCK UP?' hissed María, turning towards us.

'Sorry,' I said.

María muttered something in Spanish and then resumed her position flat against the door.

'Is it safe to go out yet?' said Ada.

'There's still footsteps outside,' said María. 'Seem to be going backwards and forwards. If I didn't know better, I'd assume they were lost.'

'Maybe it's someone else trying to break in,' I said.

'Don't be ridiculous, Tom,' said Ada.

'I still say we blast 'em,' said Waldo, reaching for his shotgun.

'Waldo, please,' said Ada.

María waved her hand at us to keep quiet and we managed to comply for a couple of minutes, at the end of which she announced, 'OK, I think it's clear.'

'You sure?' said Ada.

María looked concerned. 'Not entirely, but if I keep standing like this any more, I'm gonna get woodworm in my ear, OK?'

This seemed as good an argument for moving out as any.

'OK, let's go,' said Ada. 'Lead on.'

María led the way again as the four of us tiptoed along the dark corridor for the third time that evening. I was finding it hard to concentrate as I had now completely convinced myself that if we had to take cover again, it would quite definitely be in a room packed to the gills with venomous snakes. Waldo's assurances that he could handle them had done nothing to assuage my fears either, because as far as I could tell he hadn't packed a flamethrower in that knapsack of his, and that was the only thing I would have felt comfortable hiding behind.

Then I felt bad for even thinking that. I was a decent, animal-loving human being, wasn't I? Even scaly, poisonous reptiles with no legs that moved way faster than they had any right to deserved to be treated with respect and kindness. After all, Bertrand the ball python had got me out of a really tight spot last year, hadn't he? He'd basically saved my life. I resolved, there and then, to come back when all this was over and release everything in that penthouse into the wild. I wasn't entirely sure how I would deal with the piranhas, mind. Maybe I could build some kind of artificial river in the desert. That had to be feasible.

While going through this entirely logical sequence of thoughts in my head, I had tuned out completely, with the result that I realised far too late that María had come to a sudden halt in front of me and I couldn't stop myself from colliding with her, causing her to lurch forwards and me to trip over and fall onto the floor.

'Shit!' I said out loud.

'Sssh!' hissed three other voices simultaneously.

'OK, OK,' I said in a theatrical whisper as I staggered to my feet again. 'Could have happened to anyone.'

'No it bloody couldn't,' said Ada. 'I'm beginning to wonder why I brought you along.'

'Hey,' I said. 'You're talking to grappling hook guy here.'

'Yes, well,' said Ada, leaving the rest of her comment unspoken and hanging rudely in the air.

'Cut it out, you guys,' said María. 'I just heard the elevator arrive. We've got even more company now.'

'What?' I said. But there was no time for María to respond before Ada shoved us all into the nearest room. This was all getting absurdly complicated and I was beginning to wish I'd never got involved in this whole stupid project.

'It's Don,' said Ada, who was sneaking a look through a gap in the doorway.

'What the fuck's he doing here?' I said.

'No idea, but we need to get him out of the way,' said Ada.

'Leave him to me,' said Waldo, barging towards the front with one of his scary weapons in his hand.

'No,' said Ada. 'Got a better idea. We need a decoy.'

'Well, who's going to do that?' I said.

'Me,' said Ada, opening the door and running towards the lift.

'No, hang on—' I began, but it was too late. From my position hiding behind the door, I watched in amazement as she bowled past the oncoming figure of Don and hurled herself into the lift. Before he could do anything about it, she was on her way downwards and he was running as fast as his legs could carry him in the opposite direction towards the emergency stairs, swearing fluently in several languages as he did so.

'What's she going to do?' said María.

'Probably going to ride out into the desert and get him to follow her out there,' I said. 'Knowing the way she rides, he'll be out of our hair for some time.'

'Wow,' said María. 'OK, guys. Let's go. But don't forget, there's still someone else out there, so be wary.'

We left the room and set off once more, following María as she navigated her way through the maze. This time, we all heard the footsteps. They were cautious but steady and heading straight for us from the corridor that crossed our path just ahead. María immediately began to reverse and we all followed suit, trying every door as we went. But these were all locked.

María took a set of keys out of her pocket and tried inserting each one in turn into the lock of the nearest door to us. But nothing worked. Meanwhile the footsteps kept advancing, each step bringing us closer to our doom. Waldo took his shotgun out of his knapsack again and primed it ready for action.

Then finally, a figure emerged into the dim light in the junction between the two corridors. It was unarmed, slight of build and unexpectedly familiar.

'Put the gun down, Waldo,' I said. 'Please.'

Waldo must have picked up on the alarm in my voice, because he did exactly as I'd asked. I stepped forward and walked towards the figure ahead of us.

'Dorothy?' I said. 'Is that really you?'

Chapter 15

'Tom?' said Dorothy. 'What the hell are you doing here?'

'I could say the same to you,' I hissed.

'Excuse me,' whispered María, coming up behind me. 'Do you know this person?'

'Well, yes,' I said. 'It's Dorothy. She's my ex.'

Dorothy looked María up and down. 'Who's this?' she said.

'María,' I said. 'And this is Waldo.'

Waldo was busy rolling another joint, but gave an absent-minded wave by way of greeting.

'Well,' said Dorothy, studying Waldo with some interest. 'You have a gang. How exciting.'

'Yeah, well, it's more of a multi-disciplinary team, really,' I said.

'And what skills do you bring to the table?' said Dorothy. There was an edge to her voice which I didn't feel was entirely helpful.

'I helped them break in,' I said, trying not to sound too defensive. 'With a grappling hook.'

'No you didn't,' said Dorothy. 'They came in through the emergency door, didn't they? The one that I opened, by sliding my credit card down the gap.'

'Huh?' said María, looking up from where she was trying her keys in another door.

Dorothy gave a supercilious shrug. She'd almost certainly learned the credit card trick from Balvinder, the office intern who had started out life as a computer hacker but had moved on to physical lock-picking because it was more interesting. He insisted that it was only a hobby and he was such a pure and innocent young bloke that I actually believed him. However, in the hands of Dorothy, advanced physical hacking skills could be a dangerous weapon and I wasn't sure I approved of this new development.

'Maybe we take this conversation somewhere more private?' said María, as the door she was working on swung open. 'Just in case?'

We all followed her in and she closed the door behind us.

'I thought you were in Macau,' I said to Dorothy.

'Well, I'm not,' said Dorothy. While this was evidently true, it didn't add a lot to the conversation. 'How did you know I'd been in Macau?' she added, with a frown.

'Ali told me,' I said. 'She said it had something to do with Third Uncle.' I was hoping the fact that I was aware of this might help to prise things open. 'Doesn't he own a big casino out there or something?'

'Hang on,' came María's voice from behind me. 'Are you telling me this person is related to some kind of gambling big shot? Should we be dealing with this right now?'

I turned around and made a gesture with my hands to calm things down. 'It's OK,' I said. 'She's safe.'

'Glad you think so,' said Dorothy with a wry smile.

'I just hope I'm right,' I said, raising an eyebrow. I looked closer at Dorothy. Somewhere underneath the usual bravado, there was something else going on. She was worried. She was scared about something. She almost seemed relieved to see me.

Dorothy looked at me and sighed. 'Look, Third Uncle isn't any sort of gambling big shot. That was Ali being... well, Ali.'

'I did wonder,' I said.

'But he is fond of gambling,' continued Dorothy. 'And he's been a source of worry to the Chan family for most of my life. I mean, don't get me wrong. He's a lovely guy and he was the one who got me into mathematics, after all.'

'Right,' I said. 'I didn't realise that.'

'Anyway, from time to time,' said Dorothy, 'he gets into a bit of trouble and the family tend to send me out to fix it, because I have a better relationship with him than the rest of them.'

'Are you guys going to be much longer?' said María. 'We've got a computer to find and hack.'

'Wait,' I said. I walked back to her and whispered in her ear. 'Look,' I said, 'Dorothy can help us here, so give me another minute or two to talk her round.'

I went back to Dorothy.

'Sorry about that,' I said. 'They're a bit impatient to get cracking. What was it you were saying?'

'Well,' said Dorothy, 'the problem is that Third Uncle has got into hot water with a big shot. One who really does own a casino. Actually, not so much hot water as good old-fashioned debt. Rather a lot of it.'

'That still doesn't explain why you're here,' I said.

'I'm coming to that,' said Dorothy. 'Thing is, Third Uncle is actually pretty good at gambling and he usually does very well for himself. But he's also far too friendly and fond of a drink and he ended up somewhere playing some game he can't even remember for stupidly high stakes and the whole thing was rigged from the start.'

'Ah,' I said.

'It's not the first time either,' said Dorothy. 'But that's neither here nor there. Anyway, Third Uncle loses heavily. Very heavily. And it turns out that the only way he can stop this guy from taking his entire life savings and most of the rest of his family's life savings, plus a limb or two as a garnish, is to represent him at Robert J Merritt III's big poker night – Texas Hold 'Em style, sky's the limit. He's really good at poker, you see.'

'Bloody hell,' I said. 'Was that the plan all along?'

'I think it probably was,' said Dorothy. 'Oh, the other thing is, there's a side bet on the game, too. If Merritt loses, Third Uncle's guy also gets the deeds to El Gran Pelícano. But if Merritt wins, he gets to keep one of the other guy's places in Macau.'

'Sheesh.' I said. 'High stakes indeed. So how come you're here as well?'

'Keeping an eye on him, for one thing,' said Dorothy. 'And also to stop this game from being rigged too.'

'How do you do that?' I said.

'Merritt's data centre is somewhere up here,' said Dorothy. 'We think he has some kind of implant in his ear that gets fed by an AI thing that monitors everyone else's hand and tells him what cards to play.'

'Oh, come on,' I said. 'That's ridiculous. Isn't it?'

Dorothy shrugged. 'We'll see. Need to act fast, though. Uncle's not feeling great.'

'In what way?' I said.

'Dunno,' said Dorothy. 'Jet lag, food poisoning, might even been deliberately poisoned. Who knows? Doesn't really matter. Need to put a stop to any shenanigans before things get out of hand.'

'What happens if he doesn't win the game?' I said.

Dorothy shuddered. 'Let's not think about that,' she said.

'Better get on with it then,' I said.

'Well yes,' said Dorothy. 'Only problem is finding the data centre. This place is such a rabbit warren, though.'

'Well,' I said. 'We may be able to help you there.'

I beckoned the others over and explained our plan to Dorothy. 'So,' I said in conclusion, 'can we do a deal with you?'

'You mean, you get me into the data centre to do my stuff,' said Dorothy, 'and in return I do your stuff for you as well?'

'Basically yes,' I said.

Dorothy thought for a while. 'Yeah,' she said eventually. 'Why not?'

Now that we were confident there were none of Merritt's people in the penthouse, María took us straight to the jungle entrance. I recognised the door now, although I'd come at it from a completely different angle last time.

'OK, guys,' she said. 'This is it. Over to Waldo, I guess?'

'Are you sure there's a – what do you call it – liger in there?' said Dorothy.

'Yes,' I said. 'I've met it.'

Dorothy shook her head. 'Why does that not surprise me?' she said. 'Is it friendly?'

I turned to Waldo. 'Well?' I said. 'Is it?'

Waldo took a long drag on his joint and then patted Dorothy on the shoulder. 'Good question, my friend,' he said. 'Shall we take a look?' He grasped the door handle and turned it. The humidity of the jungle hit us all in the face like a physical force. Waldo dropped his reefer on the floor and ground it under his heel. Then he took out the pump action shotgun from his backpack and poked his head through the door. One by one we all dropped behind him as he made his way into the artificial rainforest, lit by the milky moonlight seeping in through the vast dome above us.

We'd only gone a few feet when Waldo looked round and then put his hand up for us all to stop. Jesus, had we found the liger already?

'Can one of you shut the door behind us?' hissed Waldo. We all looked at each other, then I went and gently pulled the door shut.

'Sorry!' I said.

Waldo led on into the jungle. After a while we came to a clearing and there on the ground in front of us was a sleeping big cat. It looked familiar.

'Is that her?' whispered María.

'Yep,' I said. 'That's the one.'

'Can we just sneak past her?' said María to Waldo.

'Could do,' said Waldo. 'Tad risky, though, my friend. If she wakes up and sees an evening meal strolling past, she's liable to go raid the fridge. If you see what I mean.'

'So what are you gonna do?' said María.

Waldo placed his gun on the ground before him, folded his arms and then scratched his head. 'Think I'm maybe gonna wake her up first,' he said.

'You what?' I said.

'Makes a whole load of sense,' said Waldo. 'Liger wakes up, sees her old pal, gets all cuddly. I explain you're all my best buds and she says OK, Waldo my old friend, I'll be mighty pleased to let them pass safely by.'

'And that's definitely going to work?' said Dorothy.

'Can't think of no other plan,' said Waldo.

'You're not reassuring me,' she said.

However, at that point it turned out that the liger wasn't quite as fast asleep as we'd imagined, so any discussion as to whether we should wake her up or not became moot. She pulled herself to her feet and turned towards us. She spent a second or two assessing us, as if deciding whether she could manage four in one meal or if it might be better just to start off with a couple. She could always come back for seconds.

Then she threw back her head and roared at us.

'Jesus!' I muttered.

'Don't run,' said Waldo to us, 'it'll only make things worse.'

The liger roared again.

'How much worse can it get?' said Dorothy, backing away.

Then something extraordinary happened. As we backed up towards the entrance, Waldo stepped forward and put out his hand. The big cat looked at him, put her enormous head on one side and gave out an odd whimpering sort of sound.

'There, there,' said Waldo, reaching out to pat the animal on the head. She leaned into Waldo's hand and allowed herself to be scratched. 'Been a while, eh, old girlie?' said Waldo. The liger gave out a sad whimper by way of response. I felt sorry for the poor animal, stuck up here in a crappy artificial jungle belonging to a mad billionaire with no one else for company.

The rest of us were transfixed by what was taking place before our eyes, partly in awe at how Waldo was handling the animal and partly in blind terror at the thought of what could still go horribly wrong. The thought did briefly cross my mind that, if things did go horribly wrong, the chances were that she would focus on Waldo, leaving the rest of us to make our escape. Then I felt awful for even thinking of that.

'OK, guys,' said Waldo, still focusing all his attention on the liger and keeping his eyes pointing straight ahead, 'y'all can walk on by now. Just stroll straight past her. Don't stop to pick any nice flowers you see on the way. No hesitating, no running, no jokes, no spitting. OK?'

'OK,' whispered María, leading the way past the big cat, who was now lying down on the ground again with her head in Waldo's lap, apparently waiting for someone to tickle her tummy. None of us was about to volunteer, though.

'Well, he was a good hire,' I remarked as we headed on towards the heart of the jungle.

'We're a good team,' said María. 'Only one weak link.'

'Hey,' I said.

'Someone's rumbled you quick,' said Dorothy.

'I hate you all,' I said as we arrived at the door to the operations room. 'Look, here it is. If it wasn't for me, you wouldn't even know it was here.'

'Hmmm,' said Dorothy.

'OK, you guys,' said María. 'Can we stop bickering and get on with it?'

I turned the handle and pushed the door open. The three us went in and then Dorothy pulled the door closed behind the four of us. As the door clicked shut, automatic lights flickered on and we were all temporarily blinded.

'Bloody hell,' I said. 'I wasn't expecting that.'

Dorothy ignored me, as she was staring, open-mouthed at the screen in front of us that stretched all the way across the room. It was split up into a series of individual segments, each showed either a CCTV image, a permanently evolving real-time graph or a series of scrolling numbers. In front of it was a desk, on which was a keyboard with a single glove next to it. Dorothy picked up the glove and put it on her right hand.

'Careful,' I said. 'Bad shit happens when you put on a magic glove.'

'Shut up, Tom,' said Dorothy. She pointed at one of the CCTV images and made a kind of double-tapping gesture. The image now filled the entire screen, showing a card game in progress. Then she did a kind of wave and the image separated into four individual streams, arranged in a two by two matrix, showing the hands of each of the participants in the game, with their names above each one.

'That's Third Uncle, there,' Dorothy pointed to the second stream. The name on this one read: Jimmy Chan.

'Good grief,' I said.

'Now do you believe me?' said Dorothy.

'So are you saying this is how Merritt rigs the poker game?' I said.

'Yep,' said Dorothy. 'I certainly am.'

'So are you just going to find where the feed comes in and pull the plug?' I said.

Dorothy rolled her eyes at me. 'Come on, Tom,' she said, 'that's not going to work. Think about it. If we stop the program altogether, Merritt's going to notice and get someone to fix it. Worse still, they might send someone up here to investigate.'

'So what are you going to do?' said María.

Dorothy reached into her rucksack and took out a small electronic component. She held it up for us all to admire.

'This solid state drive contains a series of faked recordings of people playing poker. All I need to do is find out where the live feeds come in and replace them with these ones. Merritt's AI will still keep feeding him advice, but it will be complete gibberish.'

'Amazing,' I said.

'Yes, that's a pretty good word for it,' said Dorothy. She had pulled the chair up to the desk and had swapped the magic glove for the keyboard and mouse.

'And you're doing this on behalf of this dubious casino owner, basically,' I said.

'It's for Third Uncle's sake,' said Dorothy. She got up, holding the hard drive in her hand, and began hunting around at the back of the big screen. Then she came back and sat down again.

'Third Uncle who's working for the dubious casino owner,' I said.

'Just let me do this, OK?' said Dorothy, waving me away. She had moved the four player streams up to the top of the big screen and was now tapping away at the keyboard and staring hard at a window which showed a load of impenetrable white text on a black background.

I stepped back and left Dorothy alone for the moment. María was looking anxious.

'Is she going to be long?' she said. 'You need to get going on your stuff as well. Don't let Ada down just because your old girlfriend has shown up.'

'It's best not to disturb her when she gets involved like this,' I said. 'We need her to focus on supporting Third Uncle first so she can help us with what we need to do. And we need to keep the game going so that we don't get trapped up here when they all come back.'

'Can we trust her?' said María.

'Of course,' I said. 'She's exceptionally irritating and obsessive at times, but she'd never double-cross us.'

'I hope you're right,' said María.

I hoped I was too. Then I had a sudden thought.

'Is everything OK with Waldo?' I said. 'We haven't heard from him for a while.'

'I'll go check,' said María. 'This place is way too bright for my eyes.'

'You sure you're OK?' I said.

'Big kitties don't scare me,' said María, heading for the door.

Once María had gone, the operations room fell silent again, the only sound being the manic clicking of Dorothy's fingers against the keyboard. I began to feel twitchy, so I started pacing up and down the length of the room. I noticed that the place wasn't quite as bright and shiny as I'd initially thought when we'd come in. There was dust accumulating in every corner, coffee stains spattering the vinyl floor and something distinctly mouldy lurking beneath the desk where Dorothy was working. I guessed that it wasn't easy to find anyone who would happily clean a place like this on a regular basis, given that the morning routine would involve making your way through a make-believe jungle guarded by a man-eating liger. Minimum wage probably wouldn't cut it.

The door opened and María came back in.

''S'OK,' she said. 'Waldo and Larry are still snuggling up together.'

'That's nice,' I said.

'Yeah,' said María. 'He was promising her he'd take her away from all this.'

'Sweet,' I said.

We were interrupted by a sudden cry of 'YESSSS!' from Dorothy. I looked up and there, below the row showing the four real participants in the poker game going on downstairs, was now a set of four decoy players.

'How come the other images are still showing?' I said.

'I'm still showing those images to anyone who's watching the live stream. Wouldn't want anyone to notice that Merritt and his chums had been swapped out.'

'But the data—' I began.

'Comes from the fake feeds, yes,' said Dorothy. She seemed inordinately pleased with herself, but I guess she deserved to be.

'Won't the AI thing, or whatever it is, realise that Merritt and his chums have been swapped out?'

'It won't be that smart.'

'You sure?'

'As sure as I need to be.'

'Cool,' I said, clapping my hands together. 'Time to get cracking on question two.'

'One moment,' said Dorothy, standing up and stretching. 'Tell me again what I'm looking for.'

'Electric motorbikes and alpacas,' I said. 'But start with the motorbikes. The alpacas appear to be collateral damage, but if you can find them as well, that would be a bonus.'

'OK, OK,' said Dorothy, 'but let's park that problem for now. So we're looking for some kind of attack software that locks up a remote piece of equipment and emails a ransom request to the owner, right?'

'Basically, yes,' I said.

'OK, let's see what we can find,' said Dorothy, pulling up a series of windows and tinkering with the keyboard. She was just getting going when her phone suddenly pinged. She picked it up and cursed.

'What's wrong?' I said.

'I think that's what's wrong,' said María, pointing up at the screen. The image from Third Uncle's stream showed him slumped forward onto the baize of the card table.

Chapter 16

'Oh my god,' I said. 'Is he dead?'

'No,' said Dorothy, not looking up from her phone. 'As I said, he's just not feeling well tonight.'

'He didn't have the enchiladas for lunch, did he?' I asked. 'And who's that messaging you?' I said.

'Oh, the casino guy from Macau,' said Dorothy.

'And he knows you're in Vegas too?' I said.

'Of course,' said Dorothy.

'Does he know you're up here?' I said.

'Of course, Tom,' said Dorothy. 'This was their plan!'

'Oh my god,' I said. 'Now we're all in cahoots with the dodgy casino guy.'

'Stop it, Tom,' said Dorothy. 'Third Uncle is unwell and they need someone to take over from him.' She stood up again and made to go towards the door.

'Hang on,' said María. 'What about the work you were going to do for us?'

'I can't do that now,' said Dorothy. 'Who else is going to take over from Third Uncle?'

'I will,' I said.

There was complete silence as we all contemplated what I had just volunteered to do.

'I'm sorry, what, Tom?' said Dorothy.

'It's obvious, isn't it?' I said. 'You're needed up here and I'm just the spare wheel. I know Bayes' theorem, too, because you told me all about it.'

Dorothy stared at me.

'Come on, Dorothy,' I said. 'You know it makes sense.'

'It makes no sense whatsoever, Tom. Are you really saying you can take on these guys at poker?' She gestured towards the screen. 'We'd be cleaned out within the first five minutes.'

'Would you do any better?' I said.

'I'd do a hell of a lot better than you would,' she said.

'Guys,' said María. 'Do you want to maybe set your personal beef aside for a moment? This ain't helping.'

Dorothy hesitated.

'Hang on,' she said. 'I've got an idea.'

'I hate it when you say that,' I said.

Dorothy was urgently tapping at her phone.

'What are you—?'

She held up her hand to forestall any attempt at conversation. There was a brief pause and then a ping as the reply came in.

'Cool,' she said. 'OK, you're in. I've explained that you're another hotshot who just happened to be in town. I've also asked them to arrange a half-hour timeout, ostensibly so that you can get changed and ready to take Third Uncle's place.'

'First problem is that I'm not a hotshot,' I said. 'Second—'.

'However,' continued Dorothy, 'what I'm also going to do in that half-hour is set up a parallel feed into a second earpiece. Which you're going to wear.'

'Hang on,' I said. 'I'm not sure I like sticking things in my ear.'

'Tough,' said Dorothy. 'Unless you really do want to try winging it? I've explained how high the stakes are, right? Figuratively *and* literally?'

I didn't say anything. I looked at María. María shrugged. Meanwhile Dorothy had already got to work and the screen was already covered in screeds of white on black verbiage that made no sense to me at all.

'Have a look in that cupboard over there,' said Dorothy. 'Bound to be some spare earpieces.'

'How do you know?' I said.

'Because they would have tested them up here, wouldn't they?' said Dorothy. 'Oh, and make sure the battery's fully charged.'

I went over to the cupboard and sure enough, there was a box on the second shelf down from the top, neatly labelled EARPIECES. I rummaged in it and found one. I gave it a clean with a tissue and looked for the 'on' switch. There was the tiniest squeak of a 'beep', and as I inserted it in my ear, a metallic voice announced that the battery level was good and that it was connecting to WiFi.

'Can you find him something to wear?' said Dorothy to María, without looking up from her programming. 'He can't go down looking like that.'

'You could say please,' said María.

'Leave her,' I said. 'She's like this when she's programming. She doesn't mean it. Well, she does, I suppose, but it's not personal.'

'OK, I'll find something,' said María. She appraised me for a moment. 'You're not that different in size from Merritt,' she added.

'Hang on,' I said. 'That reminds me. What if Jolene or Don are down there with him? They're bound to recognise me.'

'Yeah, but you're in possession of one very important little fact,' said María. 'You know about their relationship.'

'Sure,' I said, 'but chances are either Merritt knows already or he won't care one way or the other.'

'Except I also happen to know that Merritt's been seeing Jolene on the side too,' said María.

'Well, *that* is useful,' I said. 'Another good hire there,' I added.

'Of course,' said María, heading towards the door. 'Come on.'

'Oh shit,' I said. I'd just realised that I was going to have to go back through the jungle. 'Is this safe?'

'I'm sure Waldo has everything under control,' said María.

María opened the door and we stepped out into the moonlit jungle once more. This time I was more aware of the noises of small animals and I remembered the vivaria we'd found in the offices on the way here.

'It's not just the liger, is it?' I said.

'What do you mean?' said María.

'There's loads of other stuff in here that can kill us.'

'As long as you keep walking, you'll be fine.'

Something slithered across my foot. 'Yow!' I shouted, hopping around for several seconds.

'Keep quiet, Tom. You'll get all the other things excited.'

'I've never liked the word "things".'

'Tom. Just come on. Try not to look down. Trust things to get out of your way.'

'You said "things" again.'

'Tom.'

I took María's advice and kept my eyes looking firmly ahead and my feet firmly in her footsteps. Eventually we came to the clearing where we'd first found the liger. They were both still here, and she was still chilled out, lying in Waldo's arms. Waldo raised a hand in greeting.

'Yo, my friends,' he said softly.

'How are you doing?' said María.

'We're good,' said Waldo. 'I've promised her that as soon as this is all over, she's coming back to my place. This is no place for such a noble beast.'

'How are you planning to do that?' I said. The potential logistics of such an operation seemed insurmountable.

'Ain't worked it out yet,' said Waldo. 'But I sure will.'

'Cool,' I said. 'Can we go round you two? Is it safe?'

'Sure,' said Waldo with a wave of his hand. 'Be my guest.'

María and I picked our way around the sleeping animal and continued on our way towards the exit. Then we were back in the maze of corridors that led around the penthouse suite.

'OK,' said María. 'This way.' She took a left turn followed by a sharp right, before opening the fourth door on the right. 'Here we are,' she said, turning on the light. We were in a windowless room in which all four walls were covered in floor-to-ceiling cupboards behind sliding doors. María opened one to the left of the door. This contained a selection of camouflage gear and tracksuits. Then she moved along to the next one, which was host to a wide variety of business suits. She pushed the door shut and moved on again. This time she hit pay dirt. There must have been at least twenty full dinner suits, including dress shirts and bow ties, all ready for the next special occasion.

'He must go out a lot,' I said.

'Not really,' said María. 'He probably just keeps forgetting he's already got one. OK, try this one for size.'

She was holding the suit in front of me on a coat hanger. I took it from her and indicated for her to turn around so I could change.

'Seriously, Tom?' she said. 'You afraid I'm gonna laugh at you in your boxer shorts?'

'I'm very sensitive about that sort of thing,' I said.

She turned round. 'Gonna laugh anyway,' she said.

'Go ahead,' I said, taking my trousers off. I pulled on the dinner suit. It wasn't a bad fit, once I'd adjusted the waist. The shirt wasn't too bad a fit either, although it was a little on the baggy side. Finally I tied a bow tie round my neck and put the jacket on.

'OK, you can look now,' I said.

María turned around and looked at me. 'OK,' she said. 'I guess you'll do.'

I took a deep breath and prepared to leave the room.

'One more thing, Tom,' said María, holding up her hand.

'What's that?' I said.

'You've got me worried about Jolene. Even if she doesn't want Merritt to find out about her and Don, we could really do without her recognising you.' María reached into the pocket of her jeans and took out what looked like alarmingly like a flick knife. She waved it at me and motioned me to sit down. I reluctantly complied. Then with one deft movement, she opened it up and advanced towards me.

'Hold on,' I said. 'What are you up to?'

'Beard,' she said. 'Gotta go.'

'Hey, no!' I said, putting my hands up. But she was already on top of me, grabbing me by the neck and scraping away at my left cheek.

'Keep still, Tom,' she said, 'and you won't get hurt.'

I tried to protest, but all that came out was a strangulated 'Ggggggggh!' noise as she finished the left side of my face and moved towards my upper lip. Then she suddenly stopped and stood up again. My hand instinctively moved to my cheek, which stung slightly but now felt as smooth as a child's.

'OK, I'll finish now if you like,' she said.

I sighed. 'OK, OK, I take your point,' I said. 'Finish it off, then.' I would have looked really stupid turning up to the big game with a lop-sided face.

'Thank you,' said María, taking a more gentle hold of me this time and completing the job by shaving my right cheek. When she had finished, I ran my hand all over my face. She had done a pretty decent job in the end – hardly a single nick or scratch. Then she removed a hip flash from her jeans pocket, dabbed a little of the contents onto both palms and proceeded to rub my cheeks. For an instant, my entire face felt like it was on fire and then the pain subsided to a glow.

'What was that?' I said. 'Aftershave?'

'Tequila,' said María. 'Stops infection.'

There was a mirror on the wall opposite and as I stood up I caught a glimpse of the new me: suave, sophisticated and almost entirely unrecognisable. All I had to do now was pretend to be an ace poker player.

Right on cue, I heard Dorothy's voice in my ear.

'Tom? Can you hear me? Text if you can hear me.'

I got my phone out and found Dorothy's number. It had been a long time since I'd last exchanged a text with her, and that had been using a phone whose constituent parts ended up being scattered over the highway between Minsk and Chernobyl. I hesitated for a moment, then typed *Loud and clear*. María looked at me and raised an eyebrow. I mouthed the word 'Dorothy' at her.

'OK, Tom, that's good,' said Dorothy. 'Now listen. You'll need to take the emergency stairs down to the floor below. Then take the main lift down to the ground floor. Once you're down there, find somewhere safe to stash your phone.'

What? I texted.

'You won't be allowed to take your phone into the room,' said Dorothy.

What if I need to ask you about what card to play? I texted.

'I'll have already told you,' said Dorothy. 'Trust me. That's what I'm here for. Anyway, we haven't got much time, so let me continue. The poker room is at the back of the casino, but if you tell the people at the door that you're with Jimmy Chan's team, they'll take you there. Once you arrive at the room, you will be greeted by Wendell Xiang. Just tell him I sent you.'

Is he the casino owner?

'Yes,' said Dorothy. 'But don't worry about Wendell. He's quite a softie. As long as you win big, he won't lay a finger on you.'

And if I don't?

'You will,' said Dorothy. 'Just make sure you've got the earpiece switched on and stuck firmly in your ear.'

Won't they have some kind of protection against people getting information piped to them? Some sort of – what do you call it – Faraday cage?

'Tom, don't be thick,' said Dorothy. 'Merritt's using the same system himself. He can't block it. That's the beauty of this scheme.'

But won't he have realised it's not working by now?

'No, Tom,' said Dorothy, 'he'll just think it's a bit faulty, that's all. Remember, it's not supposed to work a hundred per cent of the time, because that would make people suspicious.'

Fair enough, I texted. *Anything else?*

'Don't think so,' said Dorothy. 'I'll alert you if anything goes horribly wrong. Which it won't, of course. All you need to do is keep playing and winning until Merritt throws in the towel.'

What about the other guys?

'They'll have long since bailed out,' said Dorothy. 'Neither of them has the resources to compete with either you or Merritt.'

Hang on, I typed. *Where do I get the cash from?*

'It's all done with chips,' said Dorothy. 'Wendell organises that.'

Neat.

'Oh, and one more thing,' said Dorothy. 'Don't get seduced by all the glamour, and whatever you do, promise me you won't get drunk. OK?'

I won't, I texted. I put my phone back in my pocket.

'Time to go,' I said, nodding to María. 'Can you find me the way out of here?' María led the way and we soon found ourselves back at the door to the emergency stairs.

'You OK?' she said, brushing my jacket to remove a few hairs from my beard that had collected there.

'I'm fine,' I said – and the thing was, I wasn't lying when I said this. I think it was partly the dinner suit and partly the fact that I was still alive after climbing up the side of a tall building via a rope ladder. Dorothy was back, too, and for the time being at least, she seemed to be speaking to me again, even if it was purely for the purposes of ordering me about. If I could pull this off, maybe she might begin to forgive me for what happened to the Vavasor papers.

I pulled the door closed behind me and jogged quickly along the corridor to the lift. It arrived quickly and I went straight down to the ground floor without any interruptions. As I exited the lift I noticed the restrooms just up ahead on my left. Inside, I checked that there was no one else there before heading into a single cubicle in the corner. I noted the toilet was one of those fancy old-school high-level cistern toilets with gleaming polished brass pipework that wouldn't look out of place on the *Titanic*.

I gingerly climbed onto the seat so that I could lift the lid of the cistern and peer into it. There seemed to be a dry area near the top of the far side, so I took out my phone and the roll of gaffer tape that Ada had given me earlier. Using a strip of tape, I secured the phone against the side of the cistern. As long as it didn't loosen and fall into the water, everything would be fine. I left the rest of the tape on top of the cistern and stepped back into the restroom, again checking to make sure no one had come in.

Then, I stepped back out into the lobby and turned towards the entrance to the Gran Pelícano casino.

'I'm with Jimmy Chan,' I announced to the doorman. The man looked me up and down and shook his head. 'Seriously,' I said, pointing to the walkie-talkie that the man was brandishing in his right hand. 'Check it out with Mr Xiang. He'll be out the back with the rest of the poker gang.'

The man touched his walkie-talkie and turned away from me. 'Can I speak to Mr Xiang?' I heard him say. There was a brief pause – I couldn't hear the response – and then he continued, 'Yeah... yeah...' Then, over his shoulder, he said, 'What did you say your name was?'

'Um, Winscombe,' I said. 'Tom Winscombe.'

'He says Tom Winscombe,' said the man. There was another pause, before he added, 'Yeah, kinda tall, thin, weedy, speaks with a British accent... yeah, that's the guy... OK, thank you, Mr Xiang.' The doorman switched his radio off and beckoned to me. 'Come with me,' he said.

The relief at passing the first stage of the challenge was somewhat tempered by the doorman's unflattering description of me, along with the fact that it clearly tallied with whatever Dorothy had told Wendell Xiang. I followed in the doorman's wake as he weaved his way through the crowds at the tables until we reached a door at the back of the room where a tall and completely bald Chinese guy was waiting ready to greet me. He held out his hand and I shook it. His grip was truly fearsome and I hoped that Dorothy was right about him being a real softie.

'Good evening, Mr Tom Winscombe,' he said. 'I am Wendell Xiang.' The voice was deep with the slightest hint of an American accent.

'Hi, Mr Xiang,' I said.

Look at me, I thought. Here I am, ready to play the role of Third Uncle's Second.

Chapter 17

Wendell Xiang opened the door and ushered me in to the ante room next to the poker suite. There were a number of small groups clustered together, deep in conversation and I wondered if they were the various participants discussing tactics with their supporters. There was no sign of Robert J Merritt III. Wendell pulled me to one side.

'This is all most unfortunate,' he said.

'I understand' I said. 'I trust Mr Chan is not too unwell.'

'He just needs some rest,' said Wendell. 'But in the meantime, we need a substitute or we will forfeit everything. We are already two million down.'

'I won't let you down,' I said.

'I hope you will not, Mr Winscombe,' he said. The voice was friendly, in the same way that an alligator's smile is friendly. 'Ms Chan insists that you are an excellent player,' he added, 'but I must confess I am unfamiliar with your – ah – record.'

'Ah, yes. That's because I mostly play in...' – I paused, trying to think up somewhere suitable – 'Kazakhstan,' I decided. 'The game is becoming very big there.' Christ, why did I pick Kazakhstan? I knew nothing of the place.

'Is it really?' said Wendell. His eyebrows conveyed the message that he didn't believe a single word that I was saying, but he was clearly too polite to say this out loud. 'I know Kazakhstan quite well,' he added.

Oh, bollocks.

'It's a wonderful country,' I said.

'I love the capital especially,' continued Wendell.

Jesus, what was the capital of Kazakhstan? I didn't have the faintest clue.

'It's a lovely city,' I said, ploughing on and wondering how long it was going to take to steer this heaving supertanker of a conversation towards less choppy waters.

'Which one?' he said.

'I'm sorry?'

'The old capital or the new one?'

'Um. The one with… you know… the one with the lovely… um… actually, I like both of them equally,' I said eventually. How was I to know they'd changed the capital?

Wendell gave me a knowing look that basically said, mate, you've been rumbled. I smiled back at him, doing my best wide-eyed innocent look. You may have rumbled me, mate, I was thinking, but you're stuck with me now.

'Can we talk tactics?' I said.

Wendell looked at me as if wondering whether it was even worth bothering to waste time on tactics with this useless numbskull. As things turned out, we didn't get a chance to have that discussion at all, because at that point, Robert J Merritt III and his entourage swept into the room. Merritt was in his fifties and was wearing, not unexpectedly, a remarkably similar dinner suit to the one I was wearing myself, the only difference being that he fitted it rather better than I did, being somewhat more heavily built. He was smoking a fat cigar and leading the way with Jolene and a couple of others that I vaguely recognised following in his wake. Jolene caught my eye as she passed by and she registered a minute flicker of confused recognition as she tried to work out where the hell she might have encountered me before.

Merritt's posse dropped him off at the door to the poker lounge and the rest of the players, myself included, followed him through. Whoever had designed the room had decided to go for the ambience of a high-end brothel, with a strong red and black colour scheme, outrageously comfortable chairs and lighting that emphasised the perfect green baize surface of the playing table. Cameras were positioned in various locations throughout the room, relaying streams of images to whoever might be watching outside the room, including – presumably – Dorothy upstairs in the middle of Merritt's jungle.

The only other person in the room was the croupier, a young woman dressed in an immaculate white shirt, bow tie, waistcoat and sharp-creased black trouser combo. She was carefully shuffling a pack of cards ready for use, regarding the rest of us with an inscrutable eye. I waited until the others had taken their places and then took the remaining seat for myself. As he sat down, I noticed Merritt reach up to his right ear and appear to scratch it. I instinctively felt for the device in my right ear too.

'Leave it alone, Tom,' came Dorothy's voice in my ear and my hand immediately shot down to the table. *OK, OK*, I muttered to myself. I looked at the pile of plastic chips sitting on the table in front of me. I picked one up. It was blue and had the number fifty printed on it in gold.

'Fifty thousand dollars, Tom,' came the voice in my ear. I just managed to stop myself from reacting. I glanced at the other players and noticed that they had similar sized piles in front of them. Jesus. 'Calm down,' Dorothy added. 'Just listen to what I tell you to do and everything will be fine.'

'Maybe before we get going again,' said Merritt, putting down his cigar for a moment, 'we'd better have a few introductions for our replacement player.' His voice was rich and fruity with patrician Southern undertones: the kind of guy you could imagine striding about his plantation, complaining to anyone who'd listen about the devastating impact of modern labour laws on his bottom line.

'Sure,' I said. 'I'm Tom Winscombe.'

The guy on Merritt's left scrutinised me for a moment, before saying, 'Do I know you from somewhere, boy?' He was the only one who had eschewed the conventional dinner jacket look, going instead for a full-on Nudie suit and bolo tie topped off with a preposterously large Stetson.

'I play mostly in Kazakhstan,' I said. 'Do you know the scene there?' I couldn't be bothered to come up with a better cover story, so that would have to do.

'Don't even know where it is, man,' he said with a guffaw. He looked at the other two, who both shook their heads. This was good because it meant that at least this time it wouldn't be obvious that I hadn't a clue where Kazakhstan was either.

'Willy Dunoon,' he said. 'Pleased to meet y'all. They sometimes call me Wild Cat.' I assumed this must be one of those ironic nicknames, because Wild Cat Dunoon seemed about as wild as a mildly enraged gerbil.

There was a short period of silence and then the other guy finally spoke. 'I am Novak,' he said. I stared at him, waiting for him to add to this terse statement, but it turned out that this was his entire bio.

'Cool,' I said.

Merritt took a long puff of his cigar. 'And I guess you know who I am,' he said.

'Of course, Mr Merritt, ' I said.

'Well, then,' said Merritt. He nodded to the croupier who dealt us two cards each. I had a pair of fours, diamonds and spades.

'Nice one,' said Dorothy in my ear. 'Chuck a ten on it to keep things going.'

I threw in one of my ten thousand dollar chips as nonchalantly as I could. To my left, Novak called it and then Merritt opposite did the same. Wild Cat Willy Dunoon surprised me by raising to twenty thousand.

'Ignore him,' said Dorothy. 'He's an idiot.'

Then came the flop. Three cards were laid down on the baize. Jack of hearts, two of spades and – oh, nice – four of clubs. I scratched the side of my nose.

'Careful, Tom,' said Dorothy. 'Draw them out first. And watch that scratch. Could be your tell.'

I made a point of scratching it again, just to show that it really was just an itch and not a red warning sign flashing out 'Time to fold, Winscombe's got gold'.

'Don't overdo it, Tom,' said Dorothy.

I put down a twenty. On my left, Novák folded at once. Damn. Merritt called and so did Dunoon. The pot was now standing at a hundred and ten thousand dollars. Next came the turn card, which was the Jack of clubs. Full house, fours on Jacks. My hand moved towards my pile.

'Watch it, Tom,' said Dorothy. 'Merritt's got a Jack and a two. The other Jack's still out there. If that comes out in the river, he's got you.'

What were the chances of that, though?

'One in thirty-seven,' said Dorothy, reading my mind.

Pffft. I went for it, tossing in a fifty. Merritt tapped the ash off the end of his cigar and subjected me to a penetrating stare. I blinked back at him, daring him to follow. I had no idea what Dorothy's doctored AI might be telling him about the strength of his position. He put the cigar back on his mouth and called my fifty. Dunoon bailed out.

The river card turned out to be an innocuous seven of hearts. I was home and dry with the pot currently at two hundred and ten thousand. Time to clean up. I took five of the hundred chips and chucked them in the pot. There was a brief flash of panic on Merritt's face and the end of his cigar lit up red. Then he became inscrutable again and reached for another five hundreds himself.

'Show me,' he said.

I laid down my cards. As soon as he saw what I had, Merritt hit the roof. He hurled down his own hand and squished the remnants of his cigar into the ashtray. Then he stood up and began thumping the table and gesticulating and growling something about being surrounded by 'doggone cheats' and incompetents. Novak and Wild Cat Willy Dunoon remained impassive, as if they'd seen all this many times before. While Merritt was ranting, I scooped up all the chips in the pot and added them to my pile, just in case he changed his mind.

'Is that how they play in Kazakhstan?' said Dunoon, once Merritt had completed his rant and sat down again.

'Oh yes,' I said. 'It's like the wild west out there.'

Dunoon was clearly impressed.

'OK, chill, big boy,' said Dorothy. 'Stay focused.'

Merritt lit another long fat cigar, nodded to the croupier and another round went ahead. This time Merritt folded early on and it went to a three-way face-off between Novak, Dunoon and myself, during which I – or rather Dorothy – executed an elegant bluff that forced Novak to cave unnecessarily, leaving me to face off against Dunoon's inferior hand.

My winning streak continued for the next couple of hours. With Dorothy's help, I contrived to lose a couple of hands that I could have won, just to avoid suspicion, but by the time the next break came around I had turned the situation around and I was looking at a significant profit.

'Hey,' said Dunoon, regarding his depleted chip pile with a rueful grin, 'can we have Jimmy Chan back, please?'

Merritt just looked at me suspiciously. He knew something was up, but he couldn't quite work out what it was, which was odd because it was his own system that was being used against him. Novak hadn't spoken another word since he had introduced himself and, from the look of things, he didn't have any intention of breaking his vow of silence any time soon.

We all got up to stretch our legs and we trooped off into the ante room where trays of sandwiches had been laid out for us. I hadn't realised quite how hungry I was and I immediately began stuffing my face with whatever I could grab from the buffet.

'Tom!' said Wendell, materialising in front of me. 'You play well!' He seemed surprised, so I assumed that Dorothy hadn't let him in on our little secret.

'I do my best,' I said.

'They must play hard in Kazakhstan,' said Wendell.

'Um, yes,' I said. 'Yes, they do. It's like the wild west out there.'

I almost got the impression that Wendell believed me this time.

'So what are we aiming to win?' I said.

'When our profit reaches ten million, the game ends.'

'What?'

'Ten million dollars. Then Mr Chan gets his ten per cent and his debt is cleared.'

This was the first time that I realised what kind of deal Third Uncle had got caught up in.

'He owes you a million dollars?' I said.

'Including interest, yes.'

'But that will take all night,' I said. 'And most of tomorrow too.'

'No, I do not think so,' said Wendell. 'Thanks to your good work over the last couple of hours, we are six and a half million up already. It will not take you long to get to ten million.'

'Well, I'm glad you're confident.'

'I am very confident in your abilities,' said Wendell. 'Perhaps when this is all over, you might like to consider coming to work for us? Maybe we go to Kazakhstan together?'

'I'd need to think about that,' I said. Why had I picked Kazakhstan, for god's sake? Hello, mouth, look, here comes a foot.

'Of course,' said Wendell.

Christ. Had I just received a job offer? What kind of CV would I have to provide? I guess I could get a reference from Third Uncle, though. Speaking of which…

'How is Mr Chan, by the way?' I said.

'He is feeling a little better,' said Wendell. 'In fact, he was anxious to rejoin the game. However, I told him that you were handling things superbly and that there was no need for him to rush back.'

'Did you indeed?' I said. I would have dragged him from his death bed if it had been up to me. Still, at least I had Dorothy to help me out, and with any luck, it wouldn't take me too long to accrue the last required three and a half million dollars. By then, Dorothy would also have had time to find out what was going on with Merritt's extortion scheme, so we could sort everything out and I could be back in the UK in time to help Ali and Patrice with their family plans.

I had helped myself to a slice of cake and I was halfway through eating it when Merritt appeared again and announced that the game would recommence shortly. I panicked slightly at this and stuffed the remainder of the cake into my mouth. But I tried to swallow it too quickly, with the result that I began to choke. Wendell came to my aid and immediately began to pound on my back very hard. Something shot out of my right ear and landed on the floor just underneath the advancing cowboy boot of Wild Cat Willy Dunoon. To my horror, I realised exactly what it was just in time to see the boot crush it into powder.

The good news was that at least no one was ever going to find out that I had been cheating. The very, very, very bad news indeed was that I was now no longer going to be able to cheat. I was going to have to use my own skills at poker.

Chapter 18

'Are you OK, Mr Winscombe?' said Merritt as we sat down. He lit another of his revolting cigars and leaned back in his chair.

'I'm fine now,' I said.

'Because, to my mind, you are looking a tad peaky,' he said. 'I would not be best pleased if the situation were to arise in which I had to deal with another substitution.'

'Oh no,' I said. 'I'm absolutely fine.'

'Good,' said Merritt. 'Well, let me see if I cannot claw back some of these magnificent greenbacks that you have so cleverly taken off of me this evening.'

'We'll see,' I said, trying to keep my trembling internal. This wasn't easy to do at all, as all I could see lined up for me tonight was bankruptcy and death for both myself and Third Uncle.

'Actually,' I said. 'Now that you mention it, I heard in the interval that Mr Chan was feeling somewhat better.' At least Third Uncle was a decent poker player – without my earpiece and Dorothy's help, my chances of winning what he needed had plummeted to somewhere close to zero.

'He has already removed himself from the game,' said Merritt. 'I cannot permit him to return now.'

'You sure, Bob?' said Wild Cat Willy Dunoon. 'I think I'd prefer to take my chances with old Jimmy than the new kid here.'

'So why do you think that this fine young gentleman – after such an unexpectedly felicitous run – might be volunteering the services of Mr Jimmy Chan again?' said Merritt, narrowing his eyes. 'Seems a mighty curious thing to do.'

'I was simply offering to withdraw from my temporary role and allow things to proceed as they did before,' I said. Gosh, that sounded pompous.

Merritt frowned and stared at me for a long time. 'No, son,' he said. 'You will stay here and see this through to the very end. This is a personal matter now.'

The hole that I had been staring into had just got twice as broad and ten times as deep. Robert J Merritt III was going to tear me limb from limb and then Wendell Xiang was going to have what was left for breakfast. And poor old Third Uncle Jimmy Chan was going to be the dessert. The only thing for it was to hope that the rest of the gang could stage some kind of intervention. Come to think of it, Waldo could probably do it single-handed.

I just needed to give them some kind of signal. I took my little finger, stuck it in my right ear and gave it an enthusiastic wiggle, as if earwax was the next hot commodity to invest in and I was the owner of the first commercial mine. Then I repeated the performance with my left ear, so that any observer happening to tune in at that point would be left in no doubt whatsoever that both ears were entirely bug free.

I realised that Merritt was staring at me again.

'You sure you're OK, boy?' he said.

'Me?' I said, accidentally glancing at the end of my finger. 'Never better.'

'You're a funny one, to be sure,' he said. 'And here's another particularly curious thing. My lady Jolene seems to think she knows you.'

If ever there was a time when I needed to practise my poker face, it was now. I tried as best I could, but all I managed was a weird rictus, as if I'd just realised that I'd been hypnotised into taking a hefty bite out of an onion.

'Really?' I said.

'Jolene is under the impression that you bear a truly uncanny resemblance to an individual she has had the misfortune to encounter recently.'

'I'm not sure I'm following you,' I said.

'Y'see,' said Merritt, leaning back in his chair and lighting yet another ghastly cigar, 'Jolene tells me that my man, Donald, was forced to take someone looking a lot like you for a little one-way ride out to the middle of the desert the other day. About your height. Cute British accent, too. Only this fella had a straggly beard. Now I'm not a fan of the beard myself – makes a man look slovenly and undisciplined, but that is just my personal opinion.'

'Right,' I said. The man was a lunatic, but I had to humour him for the moment. 'I couldn't agree with you m—'

'But,' he said, leaning forward suddenly and cutting me off in my tracks, 'this bearded spawn of Satan somehow escaped, leaving my poor boy Donald abandoned like the Lord Jesus Christ himself in the wilderness with only the goddamn Joshua trees for company.'

I glanced at Wild Cat Willy Dunoon and Novak. Dunoon was following the conversation with palpable interest, but it was hard to tell what Novak was thinking. He continued, as he had throughout the game, to scowl.

'But what had the guy done to deserve being taken into the desert?' I said.

'Donald caught the individual snooping around my private quarters,' said Merritt. Wild Cat Willy Dunoon let out a soft whistle and even Novak gave the slightest shake of the head. 'Truly shocking,' added Merritt.

'Gosh, that's terrible,' I said quickly.

'Ain't it just,' said Merritt. He dabbed at his cheek with the index finger of his right hand, and I couldn't stop my own index finger from mirroring the movement. 'Little nick you got there,' he said. 'Like you done shaved yourself in a bit of a hurry, boy.'

I tried hard not to gulp. The floor of the proverbial hole in front of me had now been covered with a whole truckload of venomous snakes. Merritt continued to stare at me for what seemed like six or seven hours but which in reality was probably no more than five hours at the absolute maximum. I moved my index finger to my right ear and had another good dig around.

'Well,' he said eventually. 'Shall we recommence, gentlemen?'

Novak gave the tiniest of acknowledgements and Wild Cat Willy Dunoon tipped his hat. Merritt looked at me again and raised a single eyebrow. I gave a shrug in response. Then he nodded to the croupier, who dealt us our two cards each. Now, I thought to myself. How did all that stuff with Bayes' theorem work?

Half a dozen hands later, it was desperately obvious to anyone watching that I didn't have a clue how any of this worked. My careful analysis of the behaviour of each of my three opponents, noting – for example – the varying shades of red that the end of Merritt's cigar tended to glow depending on the quality of the cards in his hand, or the number of millimetres by which Novak's scowl extended, or indeed the strange and wonderful catalogue of noises emitted by Wild Cat Willy Dunoon's mouth, came to nothing when combined with my sheer ineptitude in playing the game. When I had a winning hand, my timid bidding squandered any opportunity of making anything more than a paltry sum. When I had an obviously losing hand, I failed to realise this until I had already chucked in far too many high-value chips. Dorothy's dodgy feed was doing nothing to put Merritt off, and in fact I had a sneaking suspicion that he had abandoned it as being useless and was instead relying on his own skills. Meanwhile Novak and Wild Cat Willy Dunoon were also making a good profit from my mistakes and were both looking in a much healthier position than before.

As the croupier prepared to deal another hand, I looked at the significantly diminished pile of chips left in front of me. I hadn't really been keeping tally of how my earnings had been building up during the early part of my tenure in Third Uncle's seat, but I was pretty sure that I was now back down to somewhere close to where I had started. The gaping void was beckoning to me. If I didn't pull something out of the bag soon, we were all going to be in serious trouble.

I tried to remember all the stuff I'd trotted out the other evening about Bayes' theorem, but somehow it turned out to be way more difficult now that I was sober and confronted with a real use case. Come on, Winscombe. What would Dorothy do?

$$P(A|B) = P(B|A) \times P(A) / P(B)$$

But what did that mean again?

What if $P(A|B)$ was the probability that it would turn out that Robert Merritt had, for example, a winning hand if he took a long (say, greater than two point five seconds) drag on one of his cigars? I could use this information to decide whether or not to hang in there myself, couldn't I?

What about the other elements of the equation? $P(B|A)$ was the probability that he would take a greater than two point five second drag if he happened to have a better hand. So that was something I could estimate from observing Merritt's behaviour in the preceding hands. Of course, I hadn't actually done this in any scientific manner, because I'd been relying on Dorothy's cheat feed, but maybe I could start collecting the data now and storing it in one of those memory palaces that proper card players construct.

Now $P(A)$ was simply the probability that Merritt had a better hand than me, and that was something I should be able to work out from the cards on the table and in my hand. $P(B)$, finally, was the probability that Merritt would take a greater than two point five second drag whatever happened, and again that was something that could be studied in the field and stored in the memory palace.

There were two essential problems to this approach. First of all, I wasn't remotely convinced that I'd got the calculation of the baseline $P(A)$ right at all. It was, after all, what I'd been basing my bets on up to this point – and a fat load of use it was doing me. I had half a suspicion that I wasn't even working out the probabilities right, anyway. The other, bigger, problem was that my own memory palace, such as it was, was jerry built to start with, with shoddy foundations and showing the first tentative signs of dry rot. I had no idea what any of these numbers were, and when I made a tentative attempt to calculate a sample $P(B|A)$, I ended up with a probability of approximately 2.37, which I was pretty certain wasn't even possible.

'You OK, boy?' said Merritt, interrupting my mathematical reverie.

'Fine, fine,' I said. 'Just considering strategies.'

This was clearly very funny. 'Ho boy,' he said, looking across at Wild Cat Willy Dunoon, 'the boy's a strategiser!' Dunoon grinned in response and even Novak smirked briefly before reverting to his usual scowl. I had a strong feeling that none of these people were familiar with the work of the Reverend Thomas Bayes.

Not that it mattered, given my inability to actually put his work to any use. What was the probability of Tom winning, given that he hadn't been making any effort to gather any statistics, eh? Even I could work out that it just boiled down to the raw probability of me winning, and that was getting lower with every passing hand. Then I had an idea. Maybe it was time to bring out the Psyops. What if I could needle Merritt into making a mistake or two?

'Well,' I said, 'If I can't win by straight dealing, maybe I'll just have to indulge in some kind of EXTORTION, won't I?'

'Huh?' said Merritt, looking at me as if I was some kind of maniac.

'Maybe I'll have to BLACKMAIL you into letting me win, eh?' I said. I've never been that good at subtlety, so I rather felt that this strategy was playing to my strengths.

'What in the name of goddamn tarnation are you talking about, boy?' he said.

'Yeah,' added Dunoon. 'What are you on about, sonny?'

'Well,' I said, with what I hoped looked like an innocent smile. 'I was just thinking that extortion and blackmail are both really bad things, are they not?'

Unfortunately, Merritt did not appear to be needled by this at all, just mildly irritated.

'I am not at all sure that I get your drift here, Mr Winscombe,' he said. 'Are you perhaps suggesting that there is something untoward taking place in this mighty fine casino of mine?' He waved his arms in an expansive manner as he said this, as if to demonstrate the absolute absurdity of me calling his integrity into question.

'Good lord, no,' I said. 'But if someone were found to be using your premises to instigate and manage a global network that was indulging in cyber extortion on a planet-wide basis, raking in billions of dollars, laundered via a variety of cryptocurrencies, you'd be one hundred per cent against it, would you not?'

Merritt took out another cigar and proceeded to give me the full theatrical performance, including curtain calls and encores, of preparing it, lighting it and taking the initial few puffs to get it going. Then he leaned forward, took a long drag on the stogie and waved the glowing red tip in my face.

'Listen, sonny,' he said, in a quiet, reasonable voice. 'I don't know what the hell you think you are insinuating here, but let me tell you that, in this resplendent part of the world, we take those kinds of allegations very seriously. And I would be devastated to hear of a fine young man such as yourself having his life destroyed because of accidentally defaming the wrong individual.'

'And who would be the wrong individual in this case?' I said, going for broke.

Merritt took a long puff of his cigar and then blew a vast cloud of smoke into my face.

'Well,' he said. 'That would be for you to say out loud, wouldn't it?' He paused, and then added. 'If you were to dare.'

'Oh, I dare,' I said.

'If you dared, sonny,' he said, 'you would say it now.'

There was a silence that lasted roughly two and a half hours, while the pair of us glared at each other over the table.

'Well,' he said eventually. 'I think we'd better get this game started again.'

'We've got evidence,' I mumbled.

'Sorry, son?' said Merritt, cupping his hand behind his ear.

'Evidence,' I said, a bit louder this time.

'No, you haven't, son,' he said. 'You got nothing and you goddamn know it.' He nodded to the croupier, who shuffled the pack and dealt us all another pair of cards.

I looked at my cards. Two of clubs and three of clubs. Meh. I threw a couple of my remaining chips in: a couple of twenties. What did I have to lose? Well, actually I had a *lot* to lose, but chances are I was going to lose it anyway, so I might as well have some fun first.

Everyone called and the flop was revealed: nine of clubs, two of hearts and three of diamonds. Ooh. Two pairs already. Twos and threes, but never underestimate the low cards. I looked at what I had left in my paltry stack of chips and reckoned I could just about afford to go for a raise, assuming nothing weird happened in the later rounds of betting. I threw in a fifty. Novak folded, Merritt took a long drag on his cigar but failed to raise, and Wild Cat Willy Dunoon called as well.

The turn was the seven of clubs. Damn. I'd been hoping for another two or three. I called at fifty. Merritt raised, first his eyebrow and then his bid. We were up to sixty now. Dunoon called and we all waited for the river card. The croupier turned over the five of clubs. So all I seemed to have was two pairs. I looked at my chips. All I had left were a couple of hundreds. I picked one up and toyed with it for a moment and then tossed it over.

'Raise,' I said, looking Merritt hard in the eye. I could feel my whole body twitching in time to the beating of my heart. Surely I'd blown it now?

Merritt called and Dunoon folded. So it was just me against Merritt. Well, there was no going back. I picked up my last chip and placed it gently on the pile in the middle of the table.

'I'll see you,' I said.

Merritt carefully laid down first the nine of hearts and then the nine of spades. Three against two pairs. I'd blown it. I threw down my hand in disgust, but Merritt's reaction wasn't what I was expecting. He wasn't happy either, and he made no attempt to grab the pot. All three of them were looking at me expectantly.

Idiot. I'd failed to spot my club flush. Two and three in my hand and five, seven and nine on the table. I'd won big, by accident. Six hundred and ten thousand dollars. I was back in the game. I was still clinging on by my fingers but I felt the run of play was beginning to turn my way.

All I needed now was a run of similar accidents.

Needless to say, it didn't quite turn out that way. Instead I very soon found myself back to where I was. Or rather, around half of where I was. The run of the game had well and truly flipped right round and was cantering off into the sunset away from me. There was barely any point in continuing. A couple more hands and I would no longer be in a position to take part at all.

Then the fire alarm went off, and things suddenly became considerably less straightforward.

Chapter 19

At the sound of the deafening siren that was filling the small room, the croupier dropped everything she was holding and rushed out of the room. Merritt, Novak and Dunoon were a little slower in getting up from their chairs but they were in no less haste to depart, without a single backward glance. I waited for them to go before getting up myself, in my own time. This all seemed a little too convenient. Dorothy, after all, potentially had access to all the systems in the building, including the fire alarm, and I wouldn't have put it past her to use it to create a diversion.

I looked across at the piles of poker chips that the others had left behind. They were very tempting. Not only that, I knew that a good portion of Merritt's winnings had been gained by cheating, so what I was about to do wasn't entirely immoral. Was it?

I went to the doorway and took a look in the deserted ante room. There on the floor was a *CyberGambleCon* tote bag that someone had abandoned in their panic. I had a brief wrestling match with my conscience, knowing full well that I'd already made my mind up in the knowledge that the supposed bout was rigged from the start. I grabbed the bag and went back into the poker room. Before I had time for any second thoughts, I threw my chips into it, along with enough of Merritt's to pay off Wendell Xiang and release Third Uncle from his debt. I made a point of leaving Novak's and Dunoon's stashes alone – I didn't want to make more enemies than I needed to.

Then I sauntered out of the room, through the ante room into the casino, hoping that my bag didn't look too conspicuous. However, no one was paying any attention at all as they were all leaving the building as quickly as they could. There was no sign of Wendell Xiang at all. When I reached the door that led into the lobby of El Gran Pelícano, I snuck back towards the interior, avoiding the flow of guests streaming towards the exit and found my way to the restroom where I had secreted my phone. However, when I went to the cubicle, there was someone already in there.

'Excuse me sir,' I shouted above the din of the fire alarm. 'But there's a fire in the hotel. You need to leave now!'

There was a groan from inside the cubicle.

'Are you all right, sir?' I said, knocking hard on the door. 'You need to come out now, sir.'

There was another groan, followed by 'Oh, God,' in a voice that really did sound like someone was calling on the almighty to stop faffing about and lend a hand.

'Sir,' I said, 'I appreciate that this may be a bad time, but—'

There was another long groan and it occurred to me that I was perhaps putting my own self-interest ahead of the gentleman in the cubicle in the corner.

'Are you in need of medical attention?' I said.

This time, there was a definite attempt to form a word. I wasn't entirely sure which word, but it was a start.

'Can you spell it out?' I said.

'Urghhhhh,' came the response. It sounded disappointed. Then it spoke again. 'Eeeeee,' it said.

'OK, E. Next?'

'Nnnnnnnnnn,' was the net letter, assuming that it wasn't simply the sound of the man straining.

'Ceeeeeee,' came next, followed by 'Aishhh.'

'Sorry?' I said. 'I didn't catch that one?'

'FUCKING H!' came the desperate reply.

'OK, I'm sorry,' I said. 'So we have E – N – C – H – oh hang on, did you eat the enchiladas? Is that it?'

'Urghhhhhhhh!' came the confirmatory groan. Well, that explained everything. It also suggested that Third Uncle hadn't been poisoned – at least not deliberately. He'd probably picked the dodgy enchiladas too.

'OK, OK, I'll leave you to it then,' I said. 'Ignore the siren, by the way, I think it's just a false alarm. Do you need me to get you a doctor?'

'Nnnnnnnno jus leave malone,' said the man in the cubicle.

Fine, fine. Well, I could pick up my phone later, I guess. After all, I didn't need it to find Dorothy or the others in the penthouse. Ada was a different matter altogether, but I assumed she could look after herself. As things turned out, however, I was wrong on both counts.

It turned out that the lifts had been automatically disabled by the fire alarm going off, so I had to climb the stairs all the way there. It took me a lot longer than it should have done plus I was completely exhausted by the time I got there and in no shape to confront any threats that might be lurking in the shadows.

After I had got my breath back, I managed to find my way to the jungle again, only to see the entrance door wide open. I tiptoed over the threshold and peered into the tropical gloom. I couldn't see anything moving there, so I kept on walking towards the operations centre. Thankfully the fire alarm didn't extend this far in the building, which gave my ears a rest at least. I wished I'd had some sort of torch or at least my phone with me, because all I had to go on was the moonlight coming through the roof. After a while my eyes began to get used to the twilight and I was able to pick out a few landmarks. It wasn't long before I arrived at the clearing where I had last seen Waldo and the liger. Neither of them was there.

I carried on and very soon I was at the door to the operations room, which was also wide open. Inside, the place was deserted. That was when I began to hear voices coming from behind me towards the entrance to the jungle. I headed outside again and turned and looked back to where I had come in. I noticed a light swinging back and forth as if someone was waving a flashlight around. I quickly retreated back into the depths of the foliage and watched as two people made their way towards me.

The one with the torch was walking behind and, on closer inspection, seemed to be holding a gun in their other hand. The one in front had their hands tied together and seemed to be under some kind of duress. As they passed near me I could now see that the one walking behind was Merritt's sidekick Don. And in front of him was Ada.

'In here,' said Don, motioning Ada into the operations room. I couldn't see what was happening now, so I crept closer to the operations room and listened at the door.

'I keep telling you,' I heard Ada saying, 'I've never been here before.'

'What about the other creeps you were with?' said Don.

'I have no idea who you're talking about. It was just me up here.'

'You know who I mean. The guys who set the alarm off. The ones who've been messing around with the system here. The ones who've let Mr Merritt's liger out.'

'Mate, I have literally not got a clue what you're talking about. I know nothing whatsoever about all this stuff.'

I nudged the door open a couple of millimetres. Ada was on the left side of the room, her hands tied together, and Don was standing opposite her but side on to me, waving a gun around with reckless abandon. I calculated that I could reach him in a couple of strides. The only problem was that, if I surprised him in his present position, there was a decent chance that his gun might go off in Ada's direction. I needed to be a little more creative.

I put down the tote bag containing my poker chips and rummaged around on the jungle floor until I located a decent-sized stone. I had only one shot and I didn't trust myself to hit Don's head with it, but fortunately there was a much bigger target: that massive screen that covered the entire far wall, which was currently showing an image of the empty poker room. I eased myself around the corner of the open doorway, drew back my arm and let the rock fly. There was a smash and a shattering of glass. As Don swung his head towards the screen and I hurled myself towards him, knocking the gun out of his hand. Ada took her cue and delivered a running kick to the man's groin, forcing him to the ground in agony.

'Nice work,' I said, scooping up the gun.

'Well, thank you,' she said. 'I was wondering how I was going to get out of that one.'

'I do like to repay my debts,' I said. 'So what happened?'

'Rode out into the desert, Don followed for an hour or so, then I lost him. So I came back but he was waiting for me in the parking lot. Dragged me up all the way up here. Bit of a cock-up, really.'

'What are we going to do about Don?' I said.

'Good point,' she said. 'Can't exactly kill him in cold blood, can we? However attractive the idea might seem.'

'Got any rope?' I said.

Ada held up her bound wrists in front of me. I helped her loosen the ties and she waggled her hands around to restore her circulation. Don was still on his knees clutching his crotch. We quickly bound his hands behind him, wrapping the ends of the rope around the leg of the desk.

'Right then,' she said, 'let's get moving.'

As we left, I noticed some movement on the cracked screen on the far wall. Robert J Merritt III had returned to the poker room and began counting his chips and looking as if he could have crumbled the lot to dust in his bare hands. He was not a happy man. It was a good thing I didn't have any intention of returning to the game. We left Don and began to make our way back through the jungle to the emergency stairs.

'I have questions,' said Ada as we walked along.

'So do I,' I said, 'but you go first.'

'OK, first one,' said Ada, 'was it you who set off the fire alarm?'

'No, it was—' I almost said 'Dorothy' but then I realised that there was a whole load of stuff – starting with the presence of Dorothy in the building – that Ada was currently unaware of and that it would take time to explain, and time was something we didn't have the luxury of right now. At least, that's what I told myself. 'Actually, I'm not sure who it was,' I said. 'Maybe it was María. Or Waldo.'

Ada received this information in silence.

'OK, question two,' she said eventually, 'what's with the DJ?'

'Fancied dressing up a bit,' I said. 'And in answer to question three, the beard was getting itchy.'

'Uh-huh,' said Ada, clearly not believing a word of what I was telling her.

'Finally, what's in that tote bag?' said Ada as we walked along the corridor.

'Poker chips,' I said. 'Long story.'

'You'll have to tell me some time.'

'If we get out of here alive,' I said. 'I promise I will.'

We were almost halfway down the corridor that led to the door to the emergency stairs when I noticed the handle begin to turn, and I had to leave my questions for later.

'Shit!' said Ada, hustling me back along the corridor away from the entrance. But it was too late. Jolene burst through the door and immediately saw us scurrying away from her.

'Stop right there,' she called out.

'Tom,' hissed Ada. 'You've got Don's gun. Use it.'

'I left it back in the jungle,' I said.

'What?'

'I didn't think we'd need it,' I said.

'Jesus.'

'And besides,' I said, 'she's got an automatic.'

'Put your hands in the air,' said Jolene, right on cue.

'What do I do?' I said to Ada out of the corner of my mouth.

'Keep moving. She's not going to fire that gun,' she said. We'd been shuffling backwards as we spoke, but there was still an appreciable amount of corridor to cover before we reached anywhere that could be termed anything approaching safe. 'And when I say "Run!" turn round and zigzag your way to the door at the end, OK?'

'OK,' I said, tensing ready for action.

'RUN!' cried Ada.

We turned and fled. I heard Jolene's gun fire and a spray of bullets ricocheted off one of the walls. Just in time, we reached the door at the end, opened it and dived through, tumbling into the bar area at the interior end of the balcony pool. We continued running to the door on the far side, slamming it shut behind us as we passed through, turning the key in the lock. A blast of cold air hit us and we were back right where we'd started – on the glass balcony overlooking Las Vegas.

'Well,' said Ada.

'Well,' I said, trying very hard not to look down. The ground, as viewed through the transparent floor, was a very long way away.

'It's still here,' she said, bending down and picking up the grappling hook.

'So it is,' I said.

'Right then,' said Ada, weighing it in her hands.

'Right then,' I said.

There was a long silence.

'You're not serious?' I said.

'What's the alternative?' said Ada. 'We can't go back in there, can we? And it's only a matter of time before she gets that door open. On the other hand, they probably won't have got the glaziers in downstairs yet, will they?'

'You're mad.'

'Maybe, but I also like to stay alive.'

There was a loud banging coming from the direction of the door. I instinctively stepped away from it.

'Look,' said Ada. 'At least doing it this way around means we can place the hook safely and not rely on chucking it up into position.'

'Great,' I said. 'I love the way you like to accentuate the positives.'

'Look,' said Ada. 'I'll go first, OK?' She went to the corner where the side met the wall of the building and positioned the hook over the edge so that it was as secure as she could possibly make it. Then she unfurled the rope ladder underneath it. Finally, she took a chair and placed it just to one side of the hook.

'Don't suppose you've still got your do-it-yourself safety harness anywhere, have you?' said Ada.

'Sorry, no. I ditched it when I got changed.'

'Pity.'

There was another banging at the door. Ada took a deep breath, stood on the chair and swung her leg over the rim of the balcony onto the rope.

'OK,' she called out. 'I'm on it.'

I peered over as she slowly lowered herself down the ladder, trying very hard to focus on the very top of her head and not the ground thirty-two floors beneath her.

There was another series of blows at the door to the balcony and I had to bite my lip to stop myself from bellowing at Ada to get a move on. But she was actually making good progress and it wasn't long before she was level with the gap where the window of Room 3204 should have been. As I watched, she made a first grab at the edge and missed. She began to move her weight around in order to get the ladder swinging and I noticed in horror that the grappling hook was beginning to lift and fall as she dangled beneath it. I threw myself onto it with as much force as I could muster to keep it in place and then I saw to my relief that she had managed to get a grip on the edge and she was now slowly hauling herself into the room.

'OK,' she shouted up. 'I'm in!'

'Keep hold of the end of the rope!' I called down. I really didn't fancy repeating that manoeuvre myself. It didn't look a lot of fun.

'I'm going to tie it to this loose bit that's attached to the end of the bed,' she said. I'd forgotten that the end of the ladder had got sliced off during the earlier ascent.

'Good idea,' I said.

I slung my tote bag over my shoulder and clambered up onto the chair. Then I swung my left leg over, locating the ladder after a couple of wild attempts that succeeded only in connecting with the rarefied air. My teeth began to chatter and I really wasn't sure if it was because I was terrified or if it was simply because I was bloody cold. When I'd made the journey in the upwards direction I'd been better dressed for it. While the dinner suit endowed me with a certain Milk Tray delivery-man chic, it was hopelessly impractical from the point of view of the really important stuff, like keeping me warm. I swung my right leg over and found the rung below and then I began the descent.

I had only gone half a dozen rungs when I heard the door above me burst open and Jolene's face appeared over the railing, staring down at me with murderous intent. She seemed to be trying to decide whether to shoot me or to do something worse.

'Time for a little accident,' said Jolene before disappearing again.

There was some scuffling above me and she reappeared clutching the chair we'd used as a platform to get over the balcony wall. It was then that I realised with mounting horror what she was intending to do. First of all, she hooked one of the chair legs underneath the grappling hook. Then, with a lot of heaving, sweating and metallic clanking, she managed to use the chair leg to lever the hook free of the lip of the balcony. At this point, several quite different things happened in rapid succession.

First of all, the hook, the ladder and myself all began to plummet to earth at a speed that was significantly faster than anything I had anticipated, even in my worst imaginings. This was followed very quickly by a scraping noise and the sound of splintering wood as the bed in room 3204 began to move very fast before finding itself wedged firmly against what used to be the window. Soon after that, my downwards trajectory was brought to a sudden halt as the portion of the rope that connected me to the bed became taut, sending my legs dangling free as the rope ladder spun through a hundred and eighty degrees. A microsecond later, the grappling hook whistled past me on its continuing journey downwards, missing my ear by a matter of millimetres, before it came to a halt with another sickening jolt, swinging backwards and forwards from the end of the ladder below me.

Meanwhile, a couple of high-value chips fell out of my bag and tumbled away into the night. Assuming that they didn't happen to be underneath either of them when they landed, someone was going to get very lucky tonight. I felt this represented an unnecessary additional layer of irony because at that moment it seemed as if my own luck had well and truly run out.

Chapter 20

Somehow, I had managed to hang on to the ladder and, for a few seconds, I remained there, suspended in mid-air with my arm muscles threatening to go into spasm, while my legs flailed about beneath me trying to find anything they could get a purchase on. Eventually, my left leg located the ladder and a second or so later my right leg was there too. I carefully began to haul myself back up towards Room 3204, wondering how I was going to get in there now that the gap left by the window was now completely covered by a mattress.

'Hold on!' came Ada's muffled voice, and I could see that the mattress was beginning to shift to one side. By the time I reached the opening, there was just enough room to squeeze through, and with a helping hand from Ada, very soon, I was back on terra firma.

'I don't ever want to do anything like that ever again,' I said.

'Me neither,' said Ada, 'but we're still not out of the woods. Come on.'

We raced to the door of the room and hurtled out into the corridor. We were halfway to the lift when the door to the emergency stairs burst open behind us and Jolene came running through. For the second time in as many days, I found myself running along the corridor on the thirty-second floor, hoping that the lift was going to be clairvoyant.

'What do you reckon?' I said, looking at the lights above the lifts, which showed the one on the right ascending rapidly towards us.

'I think we're due for some luck,' said Ada. 'Look!'

As we approached the lifts, it announced its presence with a ping, the doors folded open and a familiar woman in her seventies stepped out. She wasn't trailing a suitcase this time but the expression on her face as we barged sweatily past her into the lift was pretty much identical to the previous occasion when I'd jostled with her. As the doors closed, I watched as she marched towards the oncoming Jolene, as if about to tell her off for waving her silly gun about.

Ada hammered at the button for the ground floor and eventually we began to descend.

'Told you,' said Ada.

'I don't want to rely on luck ever again,' I said. 'I've made my mind up. Everything in my life – every single little detail from now on – is going to be planned. Down to the last millisecond. Every dependency is going to have its own contingency plan. Nothing random is ever going to happen me ever again.'

Ada looked at me and tilted her head on one side.

'Really?' she said.

'Yes, really,' I said.

'Wouldn't that be a bit boring?'

'Don't care.'

The lift continued its descent and in a short while we were back on the ground floor.

'What are we going to do now?' said Ada as we came out of the lift. People were still filing back into the building after the fire alarm and there was quite a crowd mustering by the lifts.

'First thing I'm going to do,' I said, 'is go to the toilet.'

'Right,' said Ada, looking at me oddly. 'Can't you hold on until we're clear of all this?'

'I've hidden my phone in there,' I said.

'Ah. Why?'

'It's a long story. Just trust me on this.'

'OK, OK. Well, seeing as you're nipping in there,' said Ada, 'I might as well pay a visit next door.'

Ada disappeared into the Ladies' while I made my way towards the Gents'. But my pathway was blocked by a yellow plastic sign that read 'Cleaning in progress'. I peered in and saw a woman hard at work mopping the floor.

'Hello?' I said.

'Cleaning,' the woman replied, pointing to the sign.

'Yes, but I need to come in.'

'Use other toilet.'

'I can't. I need to use this one.'

The woman shrugged and went back to her mopping.

'Are you going to be long?' I said.

'Ten minutes,' she said.

'Can I give you a hand?' I said. 'I'm quite good at floors.'

She didn't even bother to respond to this, but continued her work, shaking her head as she did so.

'Mr Winscombe,' came a soft voice in my ear. 'Nice to bump into you again.'

'Ah,' I said, instinctively clutching my tote bag. 'Hi, Wendell. How fortunate. I was hoping to bump into you.'

Wendell Xiang was accompanied by a couple of guys wearing dark glasses and earpieces with ostentatious cables that disappeared into their collars. He gestured back towards the lift with an extravagant wave of his hand.

'Please lead on,' he said.

I hesitated, wondering when Ada was going to reappear. However, a none-too-gentle nudge from one of Xiang's subordinates propelled me in the direction that he had indicated. I got the strong impression that whatever resistance I might be capable of would almost certainly be useless. The gaggle of folk gathering round the lifts parted when they saw us approaching and Wendell pressed the button to go down.

We emerged into the vast underground parking lot that stretched under the hotel and into the area under the Robert J Merritt III Convention Center and began to walk towards a black Hummer that was parked on the opposite side. As we got closer, it turned its lights on full beam in our direction. Then they went off again, presumably now they had identified who we were. There was a motion in the back of the Hummer and two people were suddenly ejected from the vehicle. The first was Dorothy and the second was a rather crumpled individual wearing a dinner suit. The first thing he did was stagger over to the van in the next bay, lean on the side of it and vomit copiously on the ground. This must be Third Uncle.

'Hi!' I called out. 'Still having trouble with the enchiladas?'

Third Uncle groaned by way of response.

'So,' I said, turning to Wendell Xiang. 'Is this some kind of hostage swap thing? Poker chips for these two?' I'd been in a similar situation once before with the Belarusian mafia, where I'd ended up trading a briefcase full of explosives for Dorothy.

It turned out that Wendell Xiang didn't see it that way. Instead of answering my question, he elbowed me hard in the stomach, then as I folded up and sank to my knees, he proceeded to relieve me of my tote bag. I tried to say something in protest, but I was too winded at first to articulate anything more than a moan.

'Thought you said he was a softie,' I shouted to Dorothy. She didn't say anything in response.

'I am very courteous to winners,' said Wendell. 'Losers, not so much.' He peered into the bag. 'And yet we seem to have a small fortune in here, Mr Winscombe. I hope you didn't cheat

Mr Merritt or steal from him, because I believe that he takes such transgressions personally.'

'Good heavens, no.'

'You see,' continued Xiang. 'My problem is that if I present these fine chips to Mr Merritt's cashier, he is very likely to demand to see proof that I have earned them legitimately, and I must say that, as things currently stand, I would find it something of a struggle to do so.'

'Maybe if you don't present it all at once?' I said.

Xiang laughed. 'Ah, Mr Winscombe,' he said. 'Do you know nothing about cashflow? I pay my accountants well, but I'm not sure they would be happy if such a significant portion of my assets were to be locked up in such an illiquid form. The more pressing problem is that by doing this, you have rather sabotaged the poker game, have you not?'

'I'm sure we could come to some arrangement—'

'I very much doubt it,' said Xiang. 'And as the game is now over, I would imagine that Mr Merritt could quite reasonably argue that he has won.'

'So what are you proposing we do?' I said.

Xiang shrugged. 'I was hoping you had some ideas,' he said. 'If not, then I fear I have no further use for any of you.'

'No wait!' I said. 'There must be some way of paying back Third Unc— I mean, Jimmy Chan's debt?'

'Plus my stake money for today's game, don't forget,' said Xiang.

'How much altogether?'

'Roughly the amount in this bag,' he said, weighing it in his hand. 'But I might as well give it back to you. It's useless.'

'So are you letting us go then?' I said.

Xiang chuckled. 'If only it were that simple.' He gestured for me to join Dorothy and Third Uncle standing by the side of the van in the bay next to the Hummer.

'No, hang on,' I said, not moving. 'You can't do this.'

'Sorry,' said Xiang, 'but business is business. And this is a very bad business. Please.' He repeated the gesture and this time I complied.

'Hi,' I said to Dorothy. 'Good to see you again. Sorry about losing the earpiece and all that.'

'Idiot,' she said.

'Still, look on the bright side,' I said. 'At least neither of us had the enchiladas for lunch.' I turned to look at Third Uncle, who was bent over with his hands on his knees. He lifted his head in my direction.

'Hi,' gasped Third Uncle. His skin was the colour of an old dishcloth.

'Save your energy, Uncle,' snapped Dorothy.

'Anyway, the earpiece thing wasn't my fault,' I said. 'It was his.' I pointed towards Wendell Xiang, who was walking over to the Hummer.

'Shut up, Tom,' said Dorothy. 'I've got a plan.'

'You have?' I said.

'Someone has to have one,' said Dorothy.

Xiang banged his fist on the side of the Hummer and a couple more guys with guns emerged to join the other two standing next to Xiang.

'Have to say this doesn't look good,' I said. 'Was this part of your plan?'

'I said shut up, Tom,' said Dorothy.

'Because it looks very much to me,' I said, 'as if these guys are lining up in what I would describe as the classic firing squad formation. Does it look like that to you? Sorry if I'm getting a bit excitable and all that, but—'

A loudspeaker suddenly crackled somewhere behind us and there was a brief whine of feedback. Then a voice boomed out.

'Hold it right there,' came the voice.

Wendell Xiang peered into the darkness of the car park, then motioned to his team to lower their weapons.

'Who's that?' he called out.

'I think you know who,' said the voice. 'Your boy stole my chips.'

'It's Merritt!' I hissed to Dorothy.

'I know,' she said.

'You know?' I said. 'Are you trying to say this was part of your plan?'

'I might have tipped him off, yes,' said Dorothy.

'Jesus,' I said. I sidled along to the end of the van and risked a look. Robert Merritt was standing in the middle of the car park holding a microphone in his hand. There were four other people with him, two of whom were horribly familiar.

'He's got Don and Jolene with him,' I said. 'They're a pair of fucking psychos. Already tried to kill me several times over.'

'We'll deal with them later,' said Dorothy.

'I don't know anything about stolen chips,' Xiang was saying. 'We were in fact discussing whether this would be a good time to restart the game now that the emergency seems to have been resolved.'

'Show me the bag,' said Merritt. His amplified voice echoed around the cavernous underground space and the effect was deeply sinister, as if he was a minor demon who had been cheated out of a bunch of souls that he'd been promised.

Xiang hesitated, then gave the slightest flick of his head towards the guys from the Hummer. In one orchestrated move, they raised their weapons and fired at Merritt and his crew. Two of them fell, but Merritt, Don and Jolene had already scattered to the left and the right as a small metal disc skidded along the tarmac and came to rest underneath the Hummer.

'Woah there,' said Dorothy, spotting the disc at the same time as me. 'Let's get moving!'

Xiang's gang were distracted by the arrival of Merritt, so Dorothy grabbed hold of Third Uncle and dragged him to the front end of the van next to the wall, with me following. We managed to squeeze between the van and the wall just in time to avoid the impact of the cataclysmic explosion that launched the Hummer into the air.

'This way,' said Dorothy, heading towards a sign marked 'EXIT'. As we ran, I heard sounds of gunfire coming from the wreckage of the Hummer, but it didn't last long.

'Well, I guess Wendell's had his chips,' I said as we reached the exit.

Dorothy stared hard at me. 'Just because you're dressed up like some kind of low-rent James Bond,' she said, 'it doesn't mean you can go round making tasteless jokes like that.'

'But he is dead,' I said.

'Probably.'

'And it's basically your fault.'

'Sort of.'

'Well then.'

'Look, it's one problem we don't have any more.'

'So why can't I make a joke about it?'

There was a long silence.

'Are we going or what?' said Dorothy.

'Where are we going, though?' I said.

Footsteps began to echo around the car park – running and heading our way.

'Doesn't matter,' she said. 'We need to get away from here. Follow me.'

I followed Dorothy and Third Uncle up the stairs into the convention centre above. The place was still set out for *CyberGambleCon* but the place was in darkness and all the delegates were either tucked up in bed or out enjoying themselves.

'Now what?' I said.

'This way,' said Dorothy, pointing to our left.

'You sure?' I said.

'No, but have you got any better ideas?'

'I thought you had a plan.'

'Yes, but it only went so far.'

Meanwhile, Third Uncle had again, dropped to his knees and was retching unproductively.

'I think we need to get him to a doctor,' I said.

'He's fine,' said Dorothy. 'He's always a bit dramatic.'

'Are you sure?'

'Stop asking me if I'm sure about things. If I'm sure about something, chances are I know more about it than you do. If I'm winging it, most likely you don't know anything either. So it doesn't matter either way.'

'Bit of a generalisation,' I said. 'It's also a bit insulting.'

'It's true though.'

'Can we just get going?' I said. I grabbed Third Uncle by the scruff of the neck and lifted him up. The three of us staggered over to the side of the convention centre and tumbled into the first room we could find, just in time to avoid being spotted by whoever it was that had followed us up the stairs.

'Now what?' I said.

'We wait until they give up and go home,' said Dorothy. 'And then we go home too.'

'What happens if they come looking for us here?'

'We try to avoid that happening in the first place.' Dorothy was fiddling with her phone.

'Did you find out what was going on with Merritt's extortion game, by the way?'

'Of course. He's got this supermassive online gambling network going on which his addictive little app hooks into. When your punter back home fires up the app, it starts sending out little feelers using whatever network it can connect to and finds all your smart devices. Now, most of these come with an easily crackable default password that no one ever bothers to change because passwords are a pain in the arse, right? So it's dead easy for the app to establish a connection with all the clever shit that you have lying around your home.'

'Electric motorbikes, for example,' I said.

'Curious example to choose,' said Dorothy, 'but yes. If they can find an electric car, even better. But it tends to be white goods mainly. Smart fridges. Washing machines. Cookers. That sort of thing.'

'But what does it do with that information? Surely all those things are different? You can't just apply the same virus to each one.'

Dorothy narrowed her eyes. 'You're learning, Tom. Well done.'

God, she could be patronising sometimes. But at the same time, I preened internally at this backhanded compliment.

'Yes,' continued Dorothy. 'That's the really clever bit. The whole thing seems to be modular. It looks as if every single device has its own plug-in virus that the app delivers to it.'

'Wow,' I said. 'And if it finds something new?'

'I guess they go out and acquire one, disassemble it, figure out how it works and adapt the virus to fit it.'

'Bloody hell,' I said, suddenly thinking back to the Luxxy Duxxy factory. All those fridges and microwaves, sitting there doing nothing. 'But hang on,' I said. 'Surely someone's going to put two and two together, aren't they?'

'Well, no one's going to associate the app with stuff getting hacked. For one thing, can you imagine how many people are going to admit to their partners that they've got an addictive gambling app on their phone? In fact, I can almost imagine that the extortion deal is in two phases. Phase one says "Hey, guess what, I've just locked up all your devices, but if you send me X amount of Bitcoin I'll unlock them all for you." Phase two, on the other hand, says "Hey, if you don't send me X amount of Bitcoin, not only will I not unlock all your stuff but I'll also tell your partner what's on your phone and how much you've already gambled away."'

'Why not just do that in the first place?' I said. 'Why bother with hacking the devices?'

'Because there's no demonstration of power,' said Dorothy. 'A threat to reveal secrets to your partner is *way* more credible if the perpetrator has just bricked their hairdryer.'

'They have smart hairdryers?' I said.

'I'm willing to bet on it,' said Dorothy. 'Now keep quiet,' she added. 'I think I can hear someone outside.' She dabbed at her phone a few times and suddenly music started playing outside in the convention hall: a military band playing a Sousa march.

'Did you do that?' I hissed.

'Yes,' said Dorothy. 'Before I came down to fetch my uncle here, I installed a VNC server on several of the systems upstairs.'

'VNC?'

'Virtual Network Computing,' she said. 'Got the client on my phone. I can control most of their systems from the palm of my hand.'

'Good grief.'

'Yeah, well. Anyway, I've patched into the AV system. Should distract them for a while.' She tapped away again. 'Quite fancy some laser effects.'

Through the door, I now heard that muffled 'Pew pew' noises had been overlaid on the Sousa and I had to agree it was an inspired choice.

Dorothy reached into her bag and took out a memory stick, which she waved in my face.

'Right, Tom,' she said. 'On this drive is a copy of a file I found on their system. It's a spreadsheet that logs the details of every single device they have control over, along with the code that will unlock each one. I was going to write a quick script that would use it to release everything at once, but I didn't have time because *someone* got into difficulties in the poker game.'

'Hey, that's—' I began.

Dorothy cut me short by thrusting the memory stick into my hand. 'Look after this in case something happens to mine,' she said.

'What sort of thing?' I said.

'Anything,' said Dorothy. 'Look, I don't suppose you happen to know where the exit is?'

'Funnily enough,' I said, 'I do.' I had worked out where we were by now, and I realised that all we had to do was to find our way to the front of the hall, which was situated on our left as we came out of the door.

'OK,' said Dorothy. 'This is what I'm going to do. In thirty seconds I'm going to kill all the lights. Everything. It's going to be pitch black out there. While everyone's trying to adjust, we'll sneak past them and make our escape. Sound good?'

'OK, I guess,' I said.

Third Uncle groaned.

'Oh come on, Uncle,' said Dorothy. 'You can do this. I'm beginning to regret coming down to rescue you.'

He groaned again but managed to haul himself to his feet. We all stood there in absolute silence, waiting for the signal from Dorothy. In the dark of the room I could just about make out her hand, with the fingers counting down from five... four... three... two... one...

Chapter 21

'Go!' hissed Dorothy as we stumbled out into the convention hall. I grabbed her arm and pulled her towards where I reckoned the main entrance was, and she pulled Third Uncle along behind her. After ten seconds or so, I thought I could make out an illuminated sign that said 'Exit', and my heart began to rise.

'Come on,' I whispered to Dorothy, 'we're almost th—'

But at that point, there was a loud 'thunk' from above us and we suddenly found ourselves, quite literally, in the spotlight. We were completely exposed, right in the middle of the hall, in between a stand marketing a virtual reality system that offered punters the chance to spend an enchanting time in a fully realised virtual casino populated entirely by mythical creatures and one belonging to an energy drink company that appeared to have wandered in there by accident.

We turned round and a second 'thunk' moments later illuminated the dominating presence of Robert J Merritt III, flanked by Don on one side and Jolene on the other. They were just metres away from us. Merritt was in the process of lighting yet another of his interminable cigars. I briefly wondered if his jacket was lined with special pockets in which to secrete his stash, or if perhaps there was some kind of magic going on, like those bottomless quivers that superheroes with the gift of archery tend to use.

'Well,' said Merritt, taking a long drag. 'What do we have here? I recognise you two poker boys but the young lady in the middle is something of a mystery to me.' He turned to Don and Jolene in turn, each of whom shook their head. Then he looked back at Dorothy. 'May I enquire as to what it is that brings you to this excellent establishment of mine, young lady?' he said.

'I'm just a relation of Jimmy's,' said Dorothy.

'Are you indeed?' said Merritt. 'And your associate Mr Winscombe – I assume he is an associate of yours, and not a casual acquaintance – what brings *him* to the wonderful town of Las Vegas?'

'You'd better ask him that,' said Dorothy.

'Oh I will,' said Merritt. 'My good colleagues Donald and Jolene also have a number of highly pertinent questions for him, as they are finding his face curiously familiar.'

Judging from the looks on their faces, I felt that Merritt might have slightly exaggerated the relative importance of their desire for interrogation over their desire for my extermination – preferably using the most extended and painful mechanisms available to them. Merritt seemed to understand this, because he held up both hands briefly as if to restrain them.

'Mr Winscombe,' he said. 'The thing is, I know you're a lousy poker player. You probably play lousy poker in Kazakhstan and you play lousy poker in Las Vegas.'

'Kazakhstan?' whispered Dorothy to me.

'Yeah,' I said. 'I might have got carried away with my cover story. It's not important.'

'So the question is,' Merritt was saying, 'why are you here if it's not to play poker?' He took a couple of puffs from his cigar as if trying to decide what to say next. 'I hope you won't mind indulging me while I replay some of the events of the past few days. First of all, my delightful colleague Miss Jolene here tells me that not three days ago in this very convention hall she found herself to be lacking a team member owing to the non-appearance of our very good friend Mr Diego Estevez, and that a young gentleman of your precise height and build presented himself to her as a temporary alternative.' Jolene was nodding vigorously during this. 'Can we assume, Mr Winscombe, that the young gentleman described by Miss Jolene was indeed your very good self?'

I shrugged by way of an answer.

'I guess that I will have to take that as a response in the affirmative,' said Merritt, with some distaste. 'Be that as it may,' he continued, 'this young gentleman proved to be a reasonably able employee and was invited to return for a further day's employment, owing to the continued absence of the aforementioned Mr Diego Estevez. On the second day, however, matters took an unusual turn, did they not?' Jolene nodded even more fervently in response to this question. 'They did indeed, Mr Winscombe, they did indeed,' he said. 'Because we had what my technicians prefer to describe as an "issue" with the electrical power to our computer network and we found ourselves in sore need of a maintenance engineer in order to fix it, did we not? And who is it who steps up to the plate to help us out? Well, I'll be damned, it's our old friend. Only by the time he meets my colleague, Mr Donald, here, he is travelling under an assumed moniker: Mr Diego Rivera.'

Ha. I'd forgotten that. Dorothy was giving me a very odd look now.

'Not only that,' continued Merritt, 'but it turns out that he is about as good an electrician as you are a poker player. Which is to say, pretty goddamn terrible. So Mr Donald here begins to wonder if you may perhaps have some kind of hidden agenda, a suspicion that is only reinforced when he find you barely escaping a mauling from Thelma who likes to exercise herself in the vicinity of my special operations room.'

'Thelma?' I said.

'My liger,' said Merritt. 'She is named for my deeply beloved wife, lost to me for ten long years now.'

'She divorced him,' whispered Dorothy.

'But to continue our story,' said Merritt. 'Donald decides to take you on a small detour into our awe-inspiring desert region in order to give you some time and space to reflect on your behaviour – especially in view of the employment opportunities afforded you by those who you seem so anxious to disparage. But even then you failed to comply.'

'He was going to kill me,' I said.

'That is a small detail, Mr Winscombe,' said Merritt.

'A fairly important one,' I said.

'And yet still nothing more,' said Merritt. 'But perhaps you will permit me to conclude my narrative by considering the unusual nature of some comments that you made to me during our game of Texas Hold 'Em. If I remember correctly, your somewhat surprising insinuation was that you had evidence of some kind of criminality that you seem to think I am mixed up in. And naturally, I would be very interested to know precisely what form that evidence might perhaps take, as I regard any transgressions against the laws of man or God to be an extremely serious matter. So whatcha got, Winscombe?' Merritt amplified this last question with a forward thrust of what was left of his cigar.

'I was bluffing,' I said. 'I found no evidence for extortion or blackmail or any of the things I was accusing you of. I'm sorry.'

'Good work,' muttered Dorothy.

'Bullshit,' said Merritt. 'We know you, and your lady friend here, were in my quarters this evening. Tell me what you were doing there.'

'What are we going to do?' I whispered to Dorothy.

'Don't worry,' she said. 'I've got this.'

'What do you mean, "I've got this"?'

'I've initiated Plan D,' she said.

'Don't believe you,' I said.

'Just got to keep him talking a minute or two longer,' she said.

'Are you gonna answer my question, Mr Winscombe?' said Merritt.

'I've got it,' said Dorothy. 'Just wait one moment.' She held up her hand, crouched down and began to rummage in her bag. At this, Jolene and Don both raised their guns, but Merritt indicated that they should lower them again. After a moment, Dorothy produced a memory stick from the bag.

'Don't worry,' she whispered to me. 'This one's just got a virus on it.'

She stepped forward and tossed it in the direction of Merritt, describing a perfect parabola until he snatched it from the air in one easy movement.

'So what's this, huh?' he said.

'It's got our evidence on it,' said Dorothy. 'You're quite something, aren't you?'

'I'd watch yourself, missy, if I were you,' said Merritt.

'I can look after myself, thank you,' said Dorothy.

'Can you indeed?' said Merritt. 'Can you indeed.' He gestured to Jolene and Don, who began to walk towards us. But at this point, there was a noise at the opposite end of the hall. A door opened and María strolled in.

'Hi, everyone!' she called out, with a casual wave.

'María?' said Jolene, turning round.

'Yeah, that's me I guess,' said María. 'Good to see you again, Jolene. And you, Don. Barely recognised you with your clothes on.'

'What the fuck?' said Don, also turning round. For a moment, both Jolene and Don seemed confused as to who to go after. Then the decision became irrelevant, because someone else had just entered the room. It was Waldo.

'I was wondering where Waldo was,' I said to Dorothy.

'Was that intended to be a joke?' said Dorothy.

'It only works as a joke if you're American,' I said. 'He'd have to be Wally for it to work in Britain. But we're in America now, so—'

'Shut up, Tom,' said Dorothy.

'Sorry,' I said. 'I'm blabbering. It's just that I'm suddenly feeling ridiculously hopeful and simultaneously terrified out of my wits. I'm worried that my heart and my bowels will both burst at the same time. Also, is this your doing?'

'I might have tipped them off, yes,' said Dorothy. 'What do you think I was doing when I reached into my bag for the memory stick?'

'Well, I hope you know what you're up to,' I said.

The reason for my conflicted emotions had entered the hall at the same time as Waldo. Strolling next to him at a stately swagger, barely restrained by the lightest of cords which swung loosely between his right hand and the collar around her neck, was Thelma the liger. There was no doubt whatsoever in my mind that, should she choose to do so, Thelma would have no trouble in breaking free, but for the time being at least, she appeared to defer to Waldo in respect of all the really important decisions. This could yet all go horribly wrong. But then again, there was an outside chance that it could all go horribly right.

'OK, Thelly baby,' said Waldo. 'Easy there.' Thelma the liger came to a halt and lowered herself gracefully to the floor.

Meanwhile, Merritt had turned around and was talking to Waldo. 'Hello, Waldo,' he said. 'I see that you've reacquainted yourself with Thelma.'

'You bastard,' said Waldo.

'I'd appreciate it if you could make your language a tad less intemperate, Mr Waldo,' said Merritt, dropping his cigar on the floor and grinding the stub into the carpet. 'Unless it happens to transpire that you have some hitherto unsuspected intelligence on my parentage, that is.'

'You bastard,' repeated Waldo. 'You stole my Thelma.'

'Oh, Waldo,' said Merritt softly. 'As I'm sure you are aware, my acquisition of Thelma was the result of an entirely fair transaction between your father and myself. She was part of a full and final settlement of a position of financial embarrassment that he unfortunately found himself in, arising from debts accrued.'

'Debts accrued at your tables,' said Waldo. I wondered for a moment why he hadn't brought his arsenal of assault rifles with him, but perhaps María had persuaded him that it would all have ended very badly if he had.

'Your father chose to play,' said Merritt. A harsher tone had entered his voice. 'We are all free agents, Waldo. We are free to be virtuous or to succumb to the siren call of vice.'

'You're an evil bastard,' said Waldo.

'My position is entirely neutral,' said Merritt, spreading his arms wide. 'The position of my establishment is also entirely neutral. But enough of this absurd posturing. The thing is, Waldo, you don't scare me. Thelma is my baby now and she's not about to hurt her daddy, is she?' Merritt took a single step forward. Thelma lifted her head up and growled at him by way of response. He stepped back in alarm and Jolene and Don both edged away from him.

'Is she indeed?' said Waldo. 'Don't forget, I was the one who first trained this little kitty.' He crouched down and scratched the back of her neck. Thelma reacted to this by rolling on her back playfully.

'What have you done to her?' said Merritt, bridling.

'I've merely reawakened her true nature, Mr Merritt,' said Waldo.

Merritt stepped forward again and grabbed the lead out of Waldo's hand. 'Well, I'd like her back, please,' he said.

'Sure,' said Waldo, releasing his grip. 'She's all yours.'

As soon as Merritt took hold of her lead rope, however, Thelma sprang to her feet and gave out a full-throated roar which echoed around the cavernous convention hall. Then she took off at breakneck speed towards the exit close to where we were standing, dragging a reluctant but dogged Merritt in her wake. Dorothy grabbed hold of Third Uncle as they approached and then we parted like the Red Sea to let them through, followed seconds later by Jolene and Don, who had by now lost all interest in us now that their boss was in trouble.

Dorothy and I stared at each other for a moment, unable to quite parse what we had just witnessed. Then we turned towards María and Waldo, who were making preparations to leave the hall by the opposite exit.

'Gonna get the truck,' said Waldo. 'See if I can't head them off before any innocent bystander gets seriously hurt. Think I know where she's heading.'

'Oh my god,' said Dorothy. 'This could get very ugly.'

I looked down at Third Uncle, who was doing a passable impression of a sack of potatoes that was recovering after a heavy night out with a crate of swedes. He didn't look as if he was going anywhere fast.

'He'll be safe here,' said Dorothy, as if reading my mind.

'You sure?' I said.

'Come on,' said Dorothy.

Chapter 22

Outside the convention hall, chaos reigned. Despite the fact that it was now well past midnight, there were still plenty of punters milling around the front of the Gran Pelícano casino opposite. Everyone was looking to our right and pointing.

'Oh shit,' I said. 'Now what?'

'We head for whatever it is they're pointing at,' said Dorothy.

We made our way to the front of the crowd and saw Thelma leading the way down the strip, weaving in and out of the traffic, pulling Merritt, who was desperately trying to restrain and calm her, along behind. The two of them were closely followed by Jolene and Don, puffing along in their wake. At one point, Thelma took a detour onto the sidewalk and came close to flattening an entire phalanx of Elvis impersonators who failed to get out of the way in time. A motorbike cop overtook us and raced ahead, siren blaring, although when he caught up with Merritt and Thelma, he was clearly at a loss as to what to do. In the end he decided to ride on a little further, dismount and take up a position aiming his gun at the oncoming liger.

This wasn't an entirely sensible choice, as despite every effort by Merritt to apply the brakes, Thelma continued barrelling on towards the cop. A few metres away from him, she took off into the air, ripping the leash from Merritt's grasp and swatting the cop as she flew past him, knocking his gun out of his hand. She landed just short of his motorbike, which she proceeded to render immobile by a couple of well-directed swipes at the front tyre and fuel line respectively.

Thelma continued on her, now solo, journey down the Strip, with Jolene and Don continuing the chase while Merritt stood with his hands on his knees, wheezing ferociously. Meanwhile, the motorbike cop was staggering to his feet, flapping one hand in pain and speaking into his radio with the other.

'Officer McKenzie here,' he was saying as we caught up with him, 'request urgent backup two blocks down from the Gran Pelícano.'

'Sure, sure, we got backup to the Gran Pelícano,' came the response. 'This about the Chinese guys?'

'What Chinese guys?' said Officer McKenzie.

'The dead ones, of course,' came the response. 'The ones in the parking lot shootout.'

'What shootout?' said the cop. 'I'm talking about a big cat. On the Strip.'

'Listen, Mac,' came the voice from the radio. 'Got no time for missing cats. Best you go to the Pelícano.'

'But it's a *big* cat and my bike's trashed.'

'Don't care how big. Missing kitties ain't a priority at this time of night, bud.'

'What about my bike?'

'Can't help you there. Sorry. Make your way to the Pelícano on foot if you have to.'

I'd been transfixed listening to this.

'You OK, mate?' I said to him, but Dorothy grabbed my arm.

'Come on, Tom,' she said. 'Need to catch up with them.'

'Sorry,' I said over my shoulder to the cop, as we started jogging on down the Strip once more.

'Where do you think she's heading?' said Dorothy, after we'd been going for another couple of hundred metres.

'Waldo said he thought he knew,' I said.

'Where is Waldo anyway?' said Dorothy.

I looked at her and grinned.

'Don't,' she said.

'Also, where are Merritt and his two sidekicks?' I said.

'Shit,' said Dorothy. 'They've flagged down a cab. Look!' Up ahead of us, Merritt, Jolene and Don were piling into a taxi, which pulled away from the kerb and began to make its way down the road towards where Thelma was bounding along. Right on cue, a motorbike suddenly appeared out of nowhere, heading our way. As it drew closer, I realised it was a Harley.

'Ada!' I cried as she screeched to a halt next to us.

'Who's Ada?' said Dorothy.

'No one's mentioned Ada to you?' I said. I felt the skin on my face become unaccountably hot.

'And who's this?' said Ada, lifting the visor on her helmet.

'This is—' I began, but Dorothy interrupted me.

'I'm Dorothy,' she said, holding out her hand to Ada.

'Ah, said Ada, taking the hand and shaking it. Then she pointed at me, then her and then me again. 'That Dorothy?'

'Well, sort of,' I said. 'Although—'

'Sort of?' said Dorothy, raising an eyebrow.

'You know what I mean,' I said. Oh god, this was all so confusing.

'I'm not sure I do,' said Dorothy.

'Well, never mind what I mean. Look, anyway, this is Ada. She's the one who organised the break-in to the penthouse,' I said. 'I met her when we broke into a plastic duck factory together.'

'Not that either of us makes a habit of breaking into places, obviously,' said Ada.

Dorothy looked at Ada and then shook her head.

'I have around a hundred questions for you,' she said, 'possibly two hundred, but I guess they'll have to wait.'

'Me too,' said Ada. 'Do you want a lift? I can't take you both at the same time, but it's not far to where everyone's mustering and I can do two trips.'

I looked at Dorothy and raised an eyebrow.

'OK,' she said and climbed on.

A couple of minutes later, Ada returned.

'Kept that quiet, didn't you?' she said.

'That's not fair,' I said. 'I was going to tell you.'

'It's OK,' she said. 'I was only kidding.'

'She's worked out how the scam with your electric bike works, you know,' I said.

'Cool. You'll have to explain it all to me when we've got time. Come on, let's go.'

Life had suddenly become very complicated. Up until Dorothy had reappeared, I felt there might be something going on with Ada, but now that Dorothy was back here with me, some part of me still wondered if we could ever get together again. Of course, all of this assumed that either of them was remotely interested in me at all and this was by no means certain.

I climbed on behind Ada as she gunned the Harley down the Strip until we came to a side street where a crowd was gathering. I noticed that Merritt's cab was parked on the street next to it. She came to a halt and we both dismounted.

As we joined Dorothy, she muttered to me, 'Kept that quiet, didn't you?'

'I thought you knew,' I said. Dorothy chortled to herself, clearly very amused by my obvious embarrassment. I wasn't entirely sure if this was a good sign or not. I suspected it wasn't.

Thelma had taken up a position halfway down the street, next to Spiro's Kebab Restaurant. The door to the restaurant was shut and I could see waiters nervously peering out of the window at her. There was a gap of around twenty metres between her and the rest of the crowd. Then Merritt stepped forward.

'Hey, Thelma,' he called out. 'Whaddya think you're doing?'

Thelma growled at him by way of response. I didn't speak fluent liger, but even I could work out a rough translation. Thelma was basically saying 'Just give me the kebab now and no one gets hurt.'

'You gotta come home with me now, baby,' said Merritt, starting to walk towards her, Jolene and Don flanking him. The latter were both wearing expressions that suggested that their loyalty to their boss was already a few days past its use-by date. Behind them, the crowd was silent, trying to reach a collective decision as to whether it was going to be more entertaining if this rich out-of-towner who owned the Pelícano managed to subdue the big cat he kept as a pet, or if he ended up being eaten by that pet. It was, frankly, a tough call.

'Now would be a good time for Waldo to show up,' said Dorothy.

'No sign of him yet, though,' I said.

Merritt was now about ten metres from Thelma, still murmuring reassuring phrases to her and receiving liger abuse in return. I suddenly began to wonder if this might not end well after all. Maybe Thelma really had re-bonded with Waldo and Merritt was now surplus to her requirements. At the exact moment that this thought went through my mind, Thelma lost patience and made her move. With one magnificent leap, she was instantly fully airborne, soaring upwards and heading rapidly in his direction.

Jolene and Don glanced at each other for half a nanosecond, which was long enough for them both to agree that neither of their job descriptions included saving their boss from a marauding liger. Having made their decision, they retreated back into the crowd, abandoning him to his fate. Meanwhile, Thelma had successfully located a decent, soft landing position, right on top of Merritt. There was a single scream from him, followed by a horrible gurgle and then silence, only broken by the occasional sounds of enthusiastic chewing and breaking of minor bones.

'Oh my god,' said someone in the crowd. Then everyone began to scream and shout at once. *Wasn't someone going to do something? That poor man! Look at him! What a terrible way to go! Jesus Christ! A tiger as well! No, I think it's a lion. Definitely a tiger. Maybe a liger? What's a liger? Wasn't that the guy who owns the Pelícano? I lost a ton there once. Place is fucking rigged.*

'You OK, guys?' said María, appearing at our side.

'Yeah, we're fine,' I replied, grateful for something to distract me from looking at what was going on in front of me. It struck me that 'maul' was a remarkably small word to describe the whole panoply of the 'death by big cat' experience.

'Hi guys,' said Waldo, hot on María's heels.

'Thank god you're here,' I said. 'Can you stop her?'

Waldo took a deep breath. 'First of all,' he said, 'No way can I stop her in the middle of a meal. Second, even if I could, it's waaay too late now. And third, I think I may have told you that Robert J Merritt III is one of the biggest, most evil bastards in the world.'

'You knew she would come here, didn't you?' I said.

'She was raised on Spiro's kebabs,' said Waldo. 'They're the absolute best. I told Merritt to give her only Spiro's, but did he listen? No. Like I said, he's a bastard.'

'So what happens now?' I said.

'We wait for Thelma to finish,' said Waldo, 'and then we give her a kebab and take her away from this place. Somewhere nice.'

'Wait a moment,' I said. 'Did you say "we"?'

'Yeah,' said María, putting her arm around Waldo. If it hadn't been for the dismemberment that was going on just twenty metres away from us, it would have been quite a lovely image.

'Congratulations,' I said.

Waldo stepped forward and turned round to the crowd.

'OK, guys,' he said. 'Nothing more to see here. Time to go home and to bed.'

There was a general muttering from the crowd at this, but no one moved because there seemed to be a general expectation that tonight's show wasn't quite over yet. Waldo walked carefully over to Thelma and squatted down next to her where she was gnawing on a bone that presumably used to belong to Robert J Merritt III. He put a gentle hand on her shoulder and whispered something in her ear. Then he got up and went over to Spiro's Kebab Restaurant and tapped a couple of times on the glass of the front door. There was a short pause, then the door opened a few centimetres and one of the staff poked his head through. There was a short conversation between Waldo and the man and then the door closed again and Waldo returned to his position next to Thelma.

A minute or so later, the door to the restaurant opened again and a hand came through it holding a kebab. Thelma caught sight of this and began to get to her feet, but fortunately Waldo intercepted it just in time to avoid 'hand of waiter' joining the kebab as a garnish. Before he passed the kebab to Thelma, Waldo made her sit down again and I had the impression that they'd had the same discussion about table manners many times before when she was a cub. When Waldo was happy that Thelma was sitting down with her metaphorical serviette properly tucked in, he gave her the kebab and it was immediately obvious from the expression on her face that she found this a generally more pleasing culinary experience than Robert J Merritt III.

When she had finished her kebab, Waldo gently took hold of her leash and gave it the slightest of tugs. With great ceremony, Thelma hauled herself to her feet and shook herself as if to remove any stray crumbs from her fur. Then she and Waldo walked slowly towards us, side by side, as if she were a slightly overweight St Bernard being taken out for its evening constitutional. The spellbound crowd moved into the street to let them pass and then turned and watched, open-mouthed as Thelma bounced into the back of Waldo's truck while María and Waldo climbed into the cab. As they drove off, everyone broke into spontaneous applause, and then felt slightly bad for doing so because of what had happened to poor old Robert J Merritt III. But then again, they thought, the old bastard probably deserved it.

'Either of you guys want a lift back to the Gran Pelícano?' said Ada.

Dorothy looked at me. 'I don't know about you,' she said, 'but I could do with a walk right now, to clear my head.'

'Yeah,' I said. 'I think I could too.'

'Fair enough,' said Ada, walking over to the Harley. She donned her helmet, revved the engine and drove off into the night.

'We'd better get moving,' said Dorothy. 'The cops have finally turned up.'

A couple of policemen had indeed just arrived and were staring down at the remains of Robert J Merritt III with a mixture of fascination and revulsion.

'Any of you guys know what happened here?' one of them shouted into the dispersing crowd. What remained of the crowd gave a collective shrug. They weren't about to dob in Thelma, who had just given them so much free entertainment.

Dorothy and I walked back to El Gran Pelícano in silence. It was now in the early hours of the morning, but the casinos of Las Vegas were still running on adrenalin. When we reached the hotel, we crossed over to the convention centre and found Third Uncle just where we'd left him. Unfortunately, he wasn't alone.

'Hi,' said Jolene, waving a gun at me.

'Hi,' said Don. His gun was tucked into his waistband, but looked like it was ready to be used at a moment's notice.

'Oh god, not you two again,' I said.

'Funnily enough,' said Jolene, 'that's exactly what Don and I say whenever you two turn up. Whichever "two" it happens to be at the time.' She gestured towards Dorothy. 'Does this one know about the other one in the skin-tight leather?'

'Yes she does,' I said. 'Now can we please get on with whatever you want to do? It's been a long night.' I probably shouldn't have said this, but I was emboldened by fatigue and frankly beyond caring what happened to us now.

'OK then,' said Don, fishing for something in his pocket, 'turn around.' We did so, and then I felt him grab my wrists and bind them together with a cable tie. They did the same to Dorothy and Third Uncle, even though we both protested that there was no way he was going to put up any resistance to whatever Jolene and Don had in mind.

Don pulled out his gun and nudged me towards the exit that led down into the parking lot. Jolene did the same with Dorothy but they then realised that one of them needed to support Third Uncle.

'Can't you get this guy to stand up properly?' he said.

'Sorry,' I said. 'He had the enchiladas.'

'When, though?' said Don, looking horrified.

'They're still going through him,' I said.

Don grabbed Third Uncle by the arm and hauled him forwards. 'He's not gonna shit his pants or something, is he?' said Don.

'Might do, I guess,' I said.

'Jesus,' said Jolene. 'Just had my car valeted.'

'Well,' I said. 'That's a risk you've got to take, isn't it?'

'Oh, shut the fuck up,' said Don.

We were almost at the stairs when another figure unexpectedly emerged from them. It was Ada.

'Hi,' said Ada. 'I thought I might find you here.'

'Oh god,' said Don. 'Another one.' He turned to me. 'Wait there,' he said, walking over towards Ada with his gun in one hand and a cable tie in the other. 'OK,' he said to Ada. 'Turn round.'

'You sure you want to do this?' she said.

'Never been more certain of anything in my life,' said Don. 'You're about as much of a pain in the ass as the other two.'

'Fair enough,' said Ada, turning round and offering her wrists up for binding. She seemed way too perky about the whole thing. Don applied the ties and then he and Jolene resumed the process of moving us down to the parking lot. As we emerged from the stairs, however, we were greeted by a burst of blinding light.

'Stop right there,' came a voice. 'Drop your weapons. Las Vegas Metropolitan Police. We are armed.'

He wasn't kidding. There were about a dozen officers, arrayed in a neat formation, all pointing their guns straight at us. Apparently, the backup team weren't done investigating the Wendell Xiang multiple killing from earlier.

'Shit,' said Don.

'What should we do, Donny?' said Jolene.

'You'd better do as the man says,' sighed Don.

Jolene and Don placed their guns on the ground and held their hands up. One of the cops came forward and ushered them both towards the police van, while another came up to us and severed our plastic ties.

'Well,' said Third Uncle, pulling himself up to his full height for the first time since I'd made his acquaintance. 'That was an interesting evening.'

Chapter 23

When I finally managed to retrieve my phone from the cistern in the toilet, I noted that there were roughly two thousand WhatsApp messages from Ali reminding me of my forthcoming rescheduled appointment at what she insisted on referring to as 'The Wank Clinic'. I looked at my watch. It was three twenty in the morning. I found the app for the airline and checked when the next flight out was.

'Well, that's good,' I said.

'What is?' said Ada.

'There's a flight home this afternoon.'

'So you're not planning on hanging around?'

'No,' I said. 'I've had enough of this place for the time being.'

'Yeah, me too,' said Ada. 'You don't want to check with Dorothy?'

'Nah, she's escorting Third Uncle back to Macau. She wanted to make sure he doesn't have any obligations left to Wendell Xiang's cronies.'

'She's quite the dedicated niece, isn't she?'

'Family's important to her. Even if they drive her up the wall.'

There was a long silence between us.

'Right then,' said Ada. 'So if Dorothy really has worked out how they attacked my bike – and no, my head's not up to understanding it right now – and she can fix it, it looks like everything's tied up then. Good work all round.'

'Apart from Dolores and Steven.'

'Who?'

'The alpacas,' I said. 'I still haven't found out what's happened to them.'

'I'd forgotten about them.'

'I hadn't. They don't even belong to me, you see. That's what makes it worse.'

'Who do they belong to, then?' said Ada.

I hesitated. I still hadn't owned up to her about my connection to the Vavasors, and if I started by explaining that Dolores and Steven actually belonged to Margot Evercreech, the sometime lover of Isaac Vavasor, Ada's father, it would almost certainly open a can of worms that I didn't fancy wriggling all over me at almost half past three in the morning.

'I'll tell you one day,' I said. 'But not now. Look, can I crash on your floor for a few hours? I'm not sure I'm going to be capable of standing up for much longer.'

'You sure you don't want to ask Dorothy?'

'I don't think so,' I said. 'That might be end up being a bit complicated.'

So I ended up spending a third night on Ada Vavasor's floor.

In any other circumstances, I would have counted the ride to the airport on the back of Ada's Harley, my arms around her waist with her bag dangling from one and my own dangling from the other, as one of the more unusual experiences of my life, but after the events of the last four days it all seemed rather tame. When we got to the airport, she returned the bike to the rental company and we checked our bags in with plenty of time.

'So what are you going to do now?' I said, as we sat sipping our beers in the departure lounge bar.

'Dunno,' said Ada. 'Might try to make contact with my mother again.'

'Oh,' I said. I wasn't quite sure what reaction I was being invited to give, so I remained non-committal.

'We had a falling-out,' said Ada by way of explanation.

'Ah,' I said. 'Parents are always difficult. You should meet my father, for example. He's hopeless.'

'Really?' said Ada. 'Maybe we shouldn't be too judgmental.'

'I dunno. I think my father is one of those people who the concept of judgment was invented for.'

'Are you sure?' said Ada. 'I sometimes think there are people who just make one wrong turn and that affects everything they do from then on.'

'Why are you so upset about my father all of a sudden, anyway? What's it got to do with you?'

Ada shrugged. 'Maybe you aren't in a position to judge, is all I'm saying. Maybe you'll feel different about things in a few months when you're a father yourself.'

'That's not quite the same,' I said.

'Maybe not.'

There was a couple of minutes' silence between us while we each thought of something else to say.

'So tell me how the scam with my bike worked, then,' said Ada.

'First of all, tell me if you've got the Lucky Pelican app on your phone,' I said.

'God knows,' said Ada. 'I've got so much crap on this thing, I need to give it a good sort out.' She started scrolling through her screens. 'Hold on,' she said, 'nope... nope... nope... ooh. There it is. I knew I'd seen that logo before somewhere.' She was pointing to an app on her screen. The icon featured a generously beaked seabird that was identical to the one used in the logo for Robert J Merritt III's flagship hotel, El Gran Pelícano.

I spent the next ten minutes trying to explain to Ada, using diagrams scribbled on a paper napkin, how Merritt's gang had used the Lucky Pelican app to gain control of her electric motorbike. It had seemed much less complicated when Dorothy had described it to me, but I think I got there in the end. When I'd finished, Ada looked thoughtful.

'Doesn't explain your alpacas, though,' she said.

'I don't think anything's going to explain the alpacas,' I said. 'I'm kind of resigned to never seeing them again.'

'But how do I get my bike working again?' said Ada.

'D'oh!' I said, smacking my head theatrically. 'I forgot.' I reached into my pocket and pulled out the memory stick.

'Dorothy gave me this to look after,' I said. 'It's got details of all the devices they've taken over, along with the release codes. She was going to set up some kind of script to release them all at once, but she didn't have time. You can have it if you like – it's a copy. And I can't open it, because my laptop's knackered.' Ada took the memory stick from me and put it in her bag.

'Not sure if I'll be able to do anything with it,' said Ada, 'but I might take a look anyway. I'll email you a copy so you can pass it back to Dorothy should she need it.'

Dorothy's name hung in the air for a minute or so until Ada glanced up at the screen above our heads.

'Looks like we're boarding,' she said. 'Time to get moving.'

Seven hours later I was back in England, feeling the grey drizzle trickling down my back as I clung on behind Ada, astride another motorbike, zipping along the M25, dodging in and out of the queues of early morning commuters lined up on the big circular car park. The exhilaration of the ride almost made up for the jetlagged melancholy that was wafting over me now that I was back on my home ground. But when we arrived back at my father's place and Ada parked the bike, the feeling became so strong it was almost a physical presence.

μ, as ever, was on gatekeeping duty. Ada bent down and scratched the back of her neck and I was about to warn her that this might not be the most sensible thing to do. μ was never good with strangers. However, this time she chose to lean in as if she had known Ada for years.

'Well, you're honoured,' I said.

'Cats like me,' said Ada. 'Small ones, anyway.'

My father appeared in the doorway, looking the way he usually looked first thing in the morning – like some kind of zombie who was still trying to work out if he could be arsed breaking out of his grave.

Ada smiled up at him. 'Del!' she said. 'It *is* you, isn't it?'

'Good lord,' said my father. 'As I live and breathe. Ada!' He paused, as if worried that he'd made some kind of faux pas. 'It *is* Ada, isn't it?'

'It certainly is,' said Ada.

'You're looking good,' said my father.

'Thank you,' said Ada, beaming all over her face.

'Hang on,' I said. 'You two know each other?'

The two of them exchanged a look, as if sharing some long-forgotten secret.

'A long time ago,' said Ada eventually, 'when I was a messed-up kid trying to get away from my daft family, I ended up working at the summer festivals. That's how I met your dad.'

'Road crew, I was,' said my father. 'For all those rubbish bands at the bottom of the running order. Took Ada here under my wing. Good little worker.'

'Taught me everything I know,' said Ada.

'Everything?' I said. 'My father? What do you know about anything, Dad?'

'Cheeky sod,' he said. 'Want to come in for a cuppa?'

So we went in and the pair of them swapped war stories for half an hour about the old days when they were rigging up lighting gantries together for bands I'd never heard of while I was trying to remember what I was doing at the time, which was almost certainly a lot more boring. Eventually Ada looked at her watch and said that she really had to go. Before she climbed onto her bike, she touched my arm and said to me, 'Your dad's one of the good guys, you know.'

'Him?' I said. I found this image hard to reconcile with the dissolute wreck of a man I was currently sharing my accommodation with.

'Yes, really,' she said. Then she put her visor down and drove off. I had a feeling I'd upset her somehow, which was absurd since I was the one whose nose had just been put out, now that I'd found out that not only did I have to share my father with Ada, but I also had to share Ada with him. But perhaps I was just tired. I hadn't had a decent night's sleep for the best part of a week and everything was beginning to catch up with me. Not only that, but I still had no idea what had happened to Dolores and Steven.

I was feeling marginally livelier the next morning when I met up with Patrice and Ali outside the clinic near Harley Street in the West End of London.

'Christ,' said Ali when she saw me. 'Look what the cat just delivered.'

'Alison,' said Patrice, before enveloping me in a warm embrace. 'Good to see you again, Thomas, finally. I trust you are feeling fertile?'

'I… um, I guess so,' I said. I found her line of enquiry ever so slightly threatening, even if I don't think she didn't mean it that way. I suddenly realised what a responsibility I'd assumed on their behalf.

'Did you have a successful trip?' said Patrice.

'Sort of,' I said.

'By which I take it you didn't find your fucking alpacas,' said Ali.

'Not exactly,' I said.

'Thought so,' said Ali.

'I did bump into a friend of ours, though,' I said.

Ali frowned.

'Yep,' I said. 'Dorothy.'

'Jesus,' said Ali. 'What the fuck was she doing there? I thought she was in Macau?'

'She was,' I said. 'Then she wasn't. And now she's back there again.'

'Fuck me,' said Ali. 'What was... oh, never mind, I'll ask her. She'll probably make more sense than you.'

In this, I had to admit that Ali was probably correct. I followed them into the clinic and we were shown into a consultation room and then a nurse came in and took our details. Like every nurse the world over, she had clearly borne witness to every possible indignity that a human body had ever inflicted on its owner and had mopped up a larger quantity of unspeakable fluids that had oozed out of orifices than anyone would have imagined possible. She looked as if nothing would impress her and it was clear from the outset that I was unlikely to be in a position to change that point of view.

'So this is the putative father?' she said, looking at me with an expression of disbelief. For some reason, it was the word 'putative' that cut deep, as it seemed to be freighted with the insinuation that I was only the first of many possible options, should I turn out not to be suitable.

'Er, yes, I am,' I said. 'I am.'

She wordlessly handed me a form to sign. I scribbled my signature on it and passed it back to her. Any further verbal interaction between us was apparently no longer necessary. After some more formalities, she took my blood pressure and pulse and did the same for Patrice. Then she handed me a glass receptacle.

'Well then,' she said. For a moment, she didn't do anything else and the thought briefly crossed my mind that I was being expected to perform there and then in front of an invited audience. It was almost as if she was daring me to make the first move. After a few seconds, however, she stood up and said, 'Follow me,' and I breathed an internal sigh of relief. As I left the room, I noticed that Ali was trying very hard to suppress a snigger.

The nurse showed me into a small cubicle and explained to me that when I was finished, I should screw the top on the jar, place it on the table by the door and press the green button. She also indicated the small pile of well-thumbed magazines on the table in case I was in need of any help. Finally, if I got into any sort of difficulties, I should press the red button. It wasn't specified precisely what kind of difficulty mandated the use of the red button, but I assumed I would know what it was if and when it happened to me. Then she left me alone.

The room was bare, with white walls and a strong smell of antiseptic. There was a couch running along one side, a small wash hand basin in the far corner and a small table next to the door. I put the glass jar down on the table and then paced up and down for a minute, trying to clear my mind of intrusive thoughts. Next, it struck me that I hadn't checked to see if there were any hidden cameras, so I spent a minute or two searching the room for them. There weren't any, so I got myself ready and lay down on the couch.

I decided that I would do without the porn. This was a serious undertaking and not something to be trivialised. So I started off by thinking very hard about Ada. What would it be like with her? Did she actually wear anything under all that leather? Which colour wig would she be wearing? Or would she go without and what would I make of that?

And then out of nowhere, Dorothy appeared, elbowing Ada out of the way. Suddenly Dorothy was everywhere, Dorothy who had never truly been away from my thoughts, Dorothy my one true companion, Dorothy the only one I'd ever wanted, Dorothy,

Dorothy, Dorothy. Dorothy, oh god Dorothy, Dorothy… oh, Dorothy… I have missed you so… oh, yes, when you… and when you… and yes, oh Dorothy… oh god… oh god… yes again… once more… yes oh god oh god oh god… oh.

Oh.

Right.

A couple of minutes later I pressed the green button. The nurse returned, subjected the jar to a brief examination, grimaced as if disappointed with either the quality or the quantity, and then left the cubicle without saying another word.

'Guess what?' I said to my father when I got back home. 'You're going to be a grandfather.'

'Really?' he said. 'Are you sure?'

'Well, not really an official grandfather. Because I'm not really an official father, if you see what I mean.'

'Not entirely.'

'No, I guess this sort of thing is a bit unusual to your generation,' I said.

'Oh, is it?' said my father.

'I'm basically just the donor,' I explained.

'Right.' He appeared lost in thought for a minute. 'So, who's the lucky recipient then?' he said. 'I'm assuming it's not your new friend Ada.'

'Why not?' I said.

My father didn't say anything for a while. Then he gave a broad smile. But it wasn't a smile as if he was laughing at me. It was almost like a smile of pride.

'You don't *know*, do you?' he said.

'What don't I know?' I said.

'Oh, son,' said my father. 'You don't know how happy this makes me.'

This was getting really annoying. 'Can you stop talking in riddles, please?' I said. 'What are you going on about?'

My father took a deep breath and leaned back, staring at the ceiling.

'Let's start by saying that when I met Ada,' he said, 'she was called something different.'

'Really?' I said. 'She never told me that.'

'I doubt she needed to,' said my father. 'Or if she wanted to.'

'How odd,' I said. 'I thought she was called Ada because they're all named after mathematicians in that family, and she was named after Ada Lovelace. But what was her old name?'

'We don't mention the deadname, son,' said my father. 'It's a bit impolite.'

'What do you mean—' I began. Then I finally realised what he was saying. 'Oh I see,' I said.

'Ah, he's got it at last,' he said. He took out his pouch of tobacco and his packet of Rizlas and began to roll a cigarette. 'Worked out well for her,' he added. 'She seems happy.'

'Yes,' I said. 'I guess she does.'

I thought about this for a long time. It was going to take me a while to get used to.

'So, anyway,' said my father. 'You were going to tell me who's going to be mum.'

'Oh, yes,' I said. 'She's called Patrice. Ali's partner. That's Ali who works with Dorothy. Ali's the other mum.'

'Right,' said my father, taking a long drag on his roll-up. 'Wonderful world, isn't it, son? Wonderful, wonderful world.'

Chapter 24

The loss of Dolores and Steven meant that my days had very little structure to them, so I spent most of the rest of the day mooching around looking for something to do. At around four o'clock, however, I had a phone call. It was from Ada. I let it ring a couple of times, wondering whether or not to pick it up. Then I answered.

'Hello?' I said.

'Hi,' said Ada. 'Listen, I've found something important.' She paused. 'Tom? Are you still there?'

'Um. Yes,' I said. 'I think so.'

'Ah,' said Ada. 'He's told you, hasn't he? And you're wondering what it means for you and me, aren't you? And you're probably getting completely fixated on whether or not I'm pre-op or post-op and what *that* means for you and me, and you know what? I'm not going to tell you, because it's an impertinent question, Tom.' She paused. 'And actually, it's irrelevant as well.'

'I—'

'Sorry, that sounded a bit angry, and I'm not really angry, just not quite used to having this type of conversation yet. The other thing is I'm not even sure if it was ever going to work out for us like that anyway. I like you, Tom, don't get me wrong, but

I'm not sure I fancy you. I'm not entirely sure it's men I'm after to be honest. Still working things out. Sorry if I misled you.'

'I don't think you did,' I said.

'Anyway, glad we've got that sorted now.'

'Yes,' I said. 'I guess so.'

There was a long silence between us.

'Why Ada, though?' I said. 'I thought you didn't like being associated with mathematicians.'

'Good way to avoid people asking daft questions, though,' said Ada. There was another silence. 'So, where was I?' she said after a while. 'Oh yes. I've been looking at this spreadsheet that your friend Dorothy found – she's nice, by the way, I like her, you should get back together with her or something – and I was trying to find my electric bike on there. Haven't found the bike yet but I did come across an entry labelled "llamas". How about that?'

'Llamas?' I said. 'So, hang on, are you saying—?'

'They must have the alpacas!' said Ada. 'I mean, they think they're llamas—'

'Everyone makes that mistake,' I said.

'Well, yes, I can imagine. But anyway, it looks like they've somehow ended up with your alpacas.'

'But how?' I said. 'Unless, oh hang on. Stand by your beds, but this could start to get weird. The day before they were stolen, my dad borrowed my phone. He keeps forgetting to charge his up so he takes mine instead. Does it all the bloody time.'

'Bless him.'

'Never mind bless him. It always comes back with a load of stupid new apps downloaded onto it. Mainly crypto trading stuff that I end up having to wipe off because it's using up valuable space. And I've just remembered that I forgot to check the last couple of times. Hold on.'

I took my phone away from my ear and started scrolling through the apps. Right at the end, there it was. That bloody pelican again.

'Found it,' I said. 'Bloody thing's on my phone too.'

'What?'

'The sodding Lucky Pelican gambling app.'

'But what does that mean?' said Ada. 'It's not as if you've got any smart devices over there, is it?'

'Maybe it picked up on something inside Dolores and Steven.'

'What sort of something?'

'I dunno. Some kind of microchip. Like they have for cats and dogs.'

'Doesn't sound like the sort of thing that the Pelican app would get excited about.'

'Maybe it's a slightly more interesting microchip?' I said.

'In what way?' said Ada.

'I haven't the faintest idea. But let's imagine that they come across some device that they don't recognise and they can't find it in the shops. Maybe they send out some grunt to nick it and take it back to be analysed. In this case, the poor bugger shows up and it turns out to be a couple of alpacas.'

Ada laughed. 'So where do you think they've taken them?'

'I wonder if Merritt had any farming interests in the UK?'

'How would we find out?'

'Google? Look, let's both have a look and get back to each other in fifteen minutes. OK?'

I hung up and started searching for where Merritt's organisation might have secreted a couple of alpacas. I searched for 'Merritt+farming' and 'Merritt+agri' but nothing came up. And then I suddenly had an inspiration and there it was: Great Pelican Farm, near Tattershall in Lincolnshire. On further investigation it turned out to be a large producer of sugar beet, converted into animal feed and sold under the Merritt foods label. It had to be there. I called Ada back.

'Fancy a trip to Lincolnshire?' I said.

'You worked it out too, did you?' she said.

'If you come over here, I think I can borrow a trailer from Mad Dog McFish.'

'I'll be there first thing in the morning.'

Mad Dog McFish's ancient Land Rover Defender was, as he described it, 'self-insured'. It was also unlicensed because he took the view that, as a sovereign Freeman of the land under the jurisdiction of Magna Carta, he was not obliged to pay any form of tribute to the state, especially in the form of road tax. It had certainly never passed an MOT test either. Somehow its number plate had failed to be registered with the DVLA database and so his innovative approach to vehicle ownership had never been put to the test in court. I had grave doubts as to whether the thing would get us to Great Pelican Farm and back, but it was all I could get my hands on in a hurry. Mad Dog had lent it to me without any questions asked as I was 'a good lad and a credit to my dad', a description that – for all Ada's hard work in trying to convince me to think better of my father – I had seriously mixed feelings about.

The trailer had its eccentricities as well. The electrical connections to the Land Rover were wired up incorrectly, so that the brakes were wired up to the left-hand indicator, the left-hand indicator was wired up to the right-hand one and the right-hand indicator was wired up to the brake. This made it something of a challenge for whoever happened to get stuck behind us to work out what was going on. I did my best with hand signals, but both Ada and I were extremely grateful when we finally arrived at Great Pelican Farm without any significant incident having occurred. I brought the Land Rover to a halt a little way short of the entrance and then took a deep breath.

'Before we go in,' I said, 'there's something you need to know.'

'Oh yeah?' said Ada.

I kept staring straight ahead through the windscreen. There didn't seem to be anyone about at the farm. The place looked deserted.

'The thing is,' I said, 'you're not the only one who's been keeping secrets.'

'Go on.'

'These alpacas,' I said. 'They actually belong to a woman called Margot Evercreech. Now Margot Ever—'

'I know who Margot is,' said Ada softly.

'Oh,' I said. 'I sort of wondered if you might, what with—'

'She's my mother,' said Ada.

'Ah,' I said. 'Oh shit. I hadn't realised that. I knew she had an affair with your father, but—'

'Well, I was the outcome of that affair,' said Ada. 'But I was brought up by my father alone, with occasional help from his brothers, and yes it was every bit as weird as you might expect. My relationship with Margot was always pretty semi-detached and when I transitioned, it fell apart altogether.'

Someone was moving a tractor about up ahead.

'Actually, she was there when your father died,' I said.

'What? How do you know?'

'Because I was there too,' I said. 'So was Dorothy.'

'What?' said Ada, turning to me with eyes blazing. For a moment, she seemed to be having trouble breathing. Then she regained her composure and continued speaking, very slowly and deliberately. 'And you kept this from me?' she said. 'What else is there that you aren't telling me?'

'I was going to tell you everything, but things kept getting complicated. Anyway, I've told you now.'

'Jesus, Tom.'

'If it helps, I think she was very fond of him.'

'I don't know if it does help, to be honest.'

'There are one or two other things,' I said. 'Starting with why my dad's cat recognised you.'

'Oh god, it *was* μ!' exclaimed Ada. 'Archie and Pye's old cat! I bloody knew it. But how did your dad end up with her? What else is there, Tom?'

Up ahead, the tractor had stopped and someone was getting down from the cab.

'I'll tell you everything on the way home,' I said. 'Right now, we need to talk to this bloke.' I got out of the vehicle and closed the driver's door behind me. After some hesitation, Ada followed suit.

The man was walking towards us. He was wearing a pair of stained blue overalls, wellies and a black beanie hat. He didn't seem to be armed, which was a bonus, although he wasn't exactly smiling in welcome at us either.

'What do you want?' he called out as he approached.

'We're looking for a couple of alpacas,' I said.

'Alpacas?' he said.

'Look a bit like sheep,' I said. 'Only with a long neck. Bit aggressive sometimes, but basically friendly.'

'Oh, do you mean the llamas?' he said. 'Thank Christ for that. I was wondering what I was going to do about them.'

We followed him to an area behind a large barn where a temporary fence had been erected, and there, in the middle of it all, were Dolores and Steven, happily chomping away on the grass. They looked up briefly as I approached, gave me a vague look of recognition. Then they went back to feeding.

'Yeah, I was wondering when someone was going to come to pick them up. Bloke who dropped them off didn't seem to know much about what was going to happen about them. All sounded a bit of a cock-up. He did ask if I'd mind topping them, like, and I could keep the meat for myself and the wife, but the thing is she's become a vegan now and you know some of that stuff's really nice, I'd never have imagined it, had this lovely bean casserole last night, mind you hasn't half given me the shits, pardon my French. So no, I couldn't be responsible for killing anything like that. Even a llama.'

'Alpaca,' I said.

'You sure?' said the man, 'They look like llamas to me.'

'I'm pretty certain they're alpacas,' I said.

'So you'll be taking them back to Basingstoke then?' he said.

'Um, probably,' said Ada.

'What's going on back at HQ, anyway?' said the man. 'Heard there'd been some bother with the big boss.'

'Dunno, mate,' I said. 'No one ever tells us anything.'

'Tell me about it, mate,' said the man. 'Tell me about it.'

Ada and I led Dolores and Steven to Mad Dog's trailer. The pair of them looked at it and then at me, as if to enquire as to whether we really need to go on another road trip together.

'Sorry, guys,' I said. 'But we need to go home.'

Ada and I got in the cab and drove off in silence. After a few miles, Ada turned to me and said, 'So what else do you know about my family?'

'How long have you got?' I said.

'About as long as it takes to get back to your dad's place?'

So I told her.

Back at my father's, we drove into Mad Dog's field and unloaded the alpacas, who seemed happy to be back.

'Very soon we'll take you back to your real owner,' I said to them. 'And maybe you can give me a hand,' I said to Ada.

'I'm not sure about that,' said Ada.

'You said you wanted to make contact with her again.'

'Yes, but on my own schedule.'

'Family's important,' I said. 'Even if they make life bloody difficult for us at times.'

'We'll see,' said Ada. 'Anyway, I need to get going.'

'One moment,' I said. 'I've been thinking. About these alpacas.'

'What about them?'

'Where did Margot get them from?'

'I have no idea,' said Ada.

'Might your father have given them to her?'

'It's possible. It's the sort of thing he would have done.'

'Why?'

'Every so often he would give her little gifts to try and woo her back. But being my father, they were always slightly odd. Like a brace of alpacas, perhaps. Never did any good. She never came back to him.'

'But where would your dad have got them from?'

'God knows.'

'How long do alpacas live for?'

'Why are you asking me this? I'm not Ms Google, you know.'

'I'm just wondering if they might have once belonged to someone else in your family,' I said. 'They're not twins, are they? I mean they look as if they might come from the same litter.' This was pure speculation on my part, largely based on the fact that they were roughly the same size as each other. I turned to look at them to confirm this, but they didn't bother acknowledging me, being too busy nibbling their grass back down to its usual level.

Ada narrowed her eyes. 'Tom, where the hell are you going with this?'

'Might they have belonged to your uncles?' I said. 'Archie and Pye?'

'I... don't know,' said Ada. 'It's possible. If they had and he'd ended up inheriting them when they died, I doubt he would have wanted to hang on to them.'

'So we have a pair of alpacas, possibly twins, that might have belonged to Archimedes and Pythagoras Vavasor – perhaps two of the greatest mathematicians of our times and also twins. And each of these alpacas seem to have some kind of non-standard chip embedded in them.'

Dolores lifted her head up as I said this and gave me what I took to be an encouraging stare.

'You're way ahead of me, Tom,' said Ada. 'Or rather, you've taken a turning that I'm really not sure I want to follow you down.'

'What I'm thinking is, what if the lost mathematical secrets of the Vavasors are embedded in those chips?'

Steven was now staring at me as intently as Dolores.

'What?' said Ada. 'Are you seriously saying that every so often, my daft uncles would have uploaded their research into chips embedded in a couple of alpacas?' She thought for a moment. 'Actually,' she said. 'It would be completely on-brand. My god, that's brilliant. But how do we find out if you're right?'

'We hack into one of them,' I said.

'Say that again, Tom. Are you actually going to try and hack your way into Dolores?'

'Or maybe Steven.'

'Either way, I didn't think it was one of your skills. We need your friend Dorothy.'

'Yes, but one, she's the other side of the world right now, and two, I'm not convinced she wants anything to do with me any more.'

'Counterpoint: one, I think she actually does want to have something to do with you, and two, from what you've told me, she'd certainly do it for the Vavasors.'

'Yeah, well. Maybe.' I got out my phone, wondering how to start. Then I noticed something odd. It seemed to have connected to some kind of WiFi network.

'Hey,' I said. 'Have you seen this?'

'What?'

'It's connecting to something.'

'Go to Settings. Check the network connection.'

I did as Ada suggested and saw that I seemed to be connected to a network called 'Archimedes'.

'This isn't making any sense at all,' I said. 'I wish Dorothy was here to explain what's happening.'

'Well, there's something out here that's acting as some kind of WiFi hub thing, isn't there?' said Ada. 'And you must have connected to it before, because it's hooked into it automatically.'

'Well, I certainly haven't. I'd have remembered that.' I smacked my head. 'Oh god, I bet it's my dad. When he borrowed my phone. The 4G's a bit spotty round here and he might have gone looking for a signal and found this instead. Wouldn't have got him anywhere, though, because it's not connected to the outside world.'

'Doesn't matter, we're connected to Dolores now.'

'Or Steven.'

'Whatever. Try your browser.'

I clicked on the browser icon.

'What address are you going to put in, though?' said Ada. 'Tom? Tom, are you OK?'

I was struggling to know what to say. I hadn't needed to put any address into the browser. As soon as I called the app up it started displaying a wild swirling pattern that I recognised only too well. I showed it to Ada.

'Good grief,' she said. 'What the hell's that?'

'It's a fractal pattern,' I said.

Ada and I both stared at it for almost a minute, transfixed.

'Call her,' said Ada.

'Who?'

'You know exactly who, Tom.'

'But what time is it in Macau?'

'Doesn't matter. Call her.'

'I don't want to disturb her or anything. I mean what if—'

'Call her, Tom. Now.'

I still hesitated.

'Look,' said Ada, 'it's not yet midnight over there, so you're not going to upset anyone.'

I took a deep breath. Then I took a screen shot of the fractal and wrote her a WhatsApp message.

Hi. Found Dolores and Steven. Turns out they used to belong to Archie and Pye. Got some kind of microchip embedded. Connected to one of them. See attached image. Need help.

I disconnected the phone from the Archimedes network and checked that there was a usable 4G signal so I could send the message. There was a pause of roughly five seconds, then the phone rang.

'Tom,' came Dorothy's voice.

'Hi,' I began. 'How's Third—?'

'Don't touch anything,' she said. 'I'll be on the first plane back tomorrow.'

Then the call ended. I looked at Ada.

'Told you,' she said.

THE END

Acknowledgements

Thanks as ever to my wonderful family Gail, Mark and Rachel for their support and encouragement. Thanks once again to my publisher Pete Duncan at Farrago for putting his faith in this daft series. Thanks also to the magnificent team at Farrago for making this book the very best that it could be – especially my wonderful editor Abbie Headon and my eagle-eyed copy editor Caroline Goldsmith – and thanks to kid-ethic for yet another superb cover. Finally, thanks to the other members of the Allerton Dad's Army skittles team for introducing me to the joys of Texas Hold 'Em.

Also Available

The Truth About Archie and Pye
A Mathematical Mystery, Volume One

Something doesn't add up about Archie and Pye...

After a disastrous day at work, disillusioned junior PR executive Tom Winscombe finds himself sharing a train carriage and a dodgy Merlot with George Burgess, biographer of the Vavasor twins, mathematicians Archimedes and Pythagoras, who both died in curious circumstances a decade ago.

Burgess himself will die tonight in an equally odd manner, leaving Tom with a locked case and a lot of unanswered questions.

Join Tom and a cast of disreputable and downright dangerous characters in this witty thriller set in a murky world of murder, mystery and complex equations, involving internet conspiracy theorists, hedge fund managers, the Belarusian mafia and a cat called μ.

Also Available

A Question of Trust
A Mathematical Mystery, Volume Two

Life is not going smoothly for Tom Winscombe...

His girlfriend Dorothy has vanished, taking with her all the equipment and money of the company she ran with her friend Ali. Now Tom and Ali are forced into an awkward shared bedsit existence while they try to work out what she is up to.

Tom and Ali's investigations lead them in a host of unexpected and frankly dangerous directions, involving a pet python, an offshore stag do and an improbable application of the Fibonacci sequence. But at the end of it all, will they find Dorothy – and will she ever be able to explain what exactly is going on?

Also Available

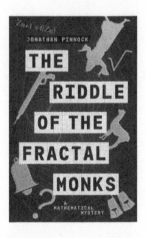

The Riddle of the Fractal Monks
A Mathematical Mystery, Volume Three

A mystery lands – literally – at Tom Winscombe's feet, and another riotous mathematical adventure begins…

Tom Winscombe and Dorothy Chan haven't managed to go on a date for some time, so it's a shame that their outing to a Promenade Concert is cut short when a mysterious cowled figure plummets from the gallery to the floor of the arena close to where they are standing. But when they find out who he was, all thoughts of romance fly out of the window.

Just who are the Fractal Monks, and what does Isaac, last of the Vavasors and custodian of the papers of famed dead mathematical geniuses Archie and Pye, want with them? How will other figures from the past also demand a slice of the action? And what other mysteries are there lurking at the bottom of the sea and at the top of mountains? The answers lie in *The Riddle of Fractal Monks*.

Also Available

Bad Day in Minsk
A Mathematical Mystery, Volume Four

High jinx in Minsk...

Tom Winscombe is having a bad day. Trapped at the top of the tallest building in Minsk while a lethal battle between several mafia factions plays out beneath him, he contemplates the sequence of events that brought him here. Starting with the botched raid on a secretive think tank; and ending up in the middle of the Chernobyl exclusion zone.

More importantly, he wonders how he's going to get out of this alive when the one person who can help is currently not speaking to him.